ROOTED IN PLACE

LATIN AMERICAN LANDSCAPES

CHRISTOPHER R. BOYER AND LISE SEDREZ

Series Editors

RICK A. LÓPEZ

ROOTED IN PLACE

Botany, Indigeneity, and Art
in the Construction of Mexican Nature, 1570–1914

THE UNIVERSITY OF
ARIZONA PRESS
TUCSON

The University of Arizona Press
www.uapress.arizona.edu

We respectfully acknowledge the University of Arizona is on the land and territories of Indigenous peoples. Today, Arizona is home to twenty-two federally recognized tribes, with Tucson being home to the O'odham and the Yaqui. The University strives to build sustainable relationships with sovereign Native Nations and Indigenous communities through education offerings, partnerships, and community service.

ISBN-13: 978-0-8165-5565-9 (hardcover)
ISBN-13: 978-0-8165-5564-2 (paperback)
ISBN-13: 978-0-8165-5591-8 (ebook)

Cover design by Leigh McDonald
Cover art from the Codex Cruz-Badiano, courtesy of Instituto Nacional de Antropología e Historia, Mexico City
Typeset by Leigh McDonald in Adobe Jenson Pro 10.5/11 and Waters Titling Pro (display)

Publication of this book is made possible in part by support from The Trustees of Amherst College.

Library of Congress Control Number: 2025003807

Printed in the United States of America
♾ This paper meets the requirements of ANSI/NISO Z39.48-1992 (Permanence of Paper).

CONTENTS

ILLUSTRATIONS

FIGURES

MAPS

ACKNOWLEDGMENTS

The intellectual journey that led to this book would have been impossible without the many individuals who supported me along the way. John Soluri, Emily Wakild, Myrna Santiago, Cynthia Radding, Lise Sedrez, and Chris Boyer joined me to think together for several days back in 2007, at a time when I was first finding my bearings within the field of Latin American environmental history. For opportunities to share my research during its initial stages, I want to thank Raymond Craib, Kate Fady, Brian Ogilvie, and Gilbert Joseph for inviting me to share some of my early findings at Cornell University, the New York Botanical Garden, the Arthur F. Kinney Center for Interdisciplinary Renaissance Studies at UMass, and Yale University, respectively.

I owe an enormous debt to the many people who assisted me in my research, and those who gave me feedback on all or part of the manuscript. This includes Antonio González Bueno for orienting me within botanical archives in Madrid; Robbin C. Moran of the William and Lynda Steere Herbarium of the New York Botanical Garden for showing me the intricacies of how herbarium collections operate; Eric Van Young for generously sharing documents related to Lucas Alamán; Mary Coffey and Dafne Cruz Porchini for help with image permissions; and Jesse Barba for assisting with data analysis. Several former students, most notably Rachel Meketon, provided invaluable assistance with paleography and transcription.

This book owes a great deal to the incredible work of archivists and curators in Mexico, the United States, and Europe, with special appreciation to Carrie Roy at the Hunt Institute for Botanical Documentation and Mike Kelly at the Amherst College Archives and Special Collections. Thanks to Mackenzie Cooley, Rob Alegre, Michael Dietrich, Chris Boyer, and the students in my Environmental History of Latin America course (fall 2024) for insightful feedback and suggestions on all or part the book manuscript. And thank you, also, to the folks at the University of Arizona Press, especially Kristen Buckles, Elizabeth Wilder, Amanda Krause, Heather Jacobson, and the anonymous reviewers, for their help improving the text and making this book a reality.

This project would not have been possible without the generous funding and institutional support of Amherst College, including its Faculty Research Award program and the Office of the Provost. At Amherst, I also want to thank my students and colleagues in history, environmental studies, and Latinx and Latin American studies, especially Anna Martini (who team-taught an interdisciplinary environmental studies course with me and patiently answered numerous questions about geology and hydrology) and Nicola Courtright (for serving as a mentor for the past thirty-plus years, and team-teaching a first-year seminar course with me about nature).

My commitment to scholarly research also owes a great deal to relationships beyond the academy, starting with my parents Alejandro López and Yolanda Consuelo Marestein López (d. 2021), who taught me to love the outdoors and nurtured my fascination with the natural world. For keeping me centered and for their unwavering support, I want to thank my extended family in El Paso and New England, especially Sharon Church, Manny Rivera Sr., Bob Griffin, Gladys Rivera Griffin, Manly Church, and my Behnke compadres. For their companionship while exploring the wonders of the natural world I want to thank my fellow Pittsburg *monteros*, with special shout-outs to Jan Dizard, Manny Rivera Jr., Aaron Keeler, and John Ketcham for introducing me to the backwoods of New Hampshire. I am grateful to my children Amaya and Diego for filling my days with joy. And most of all I want to thank my wife, Amanda, the love of my life to whom I owe more than I can put into words.

ROOTED IN PLACE

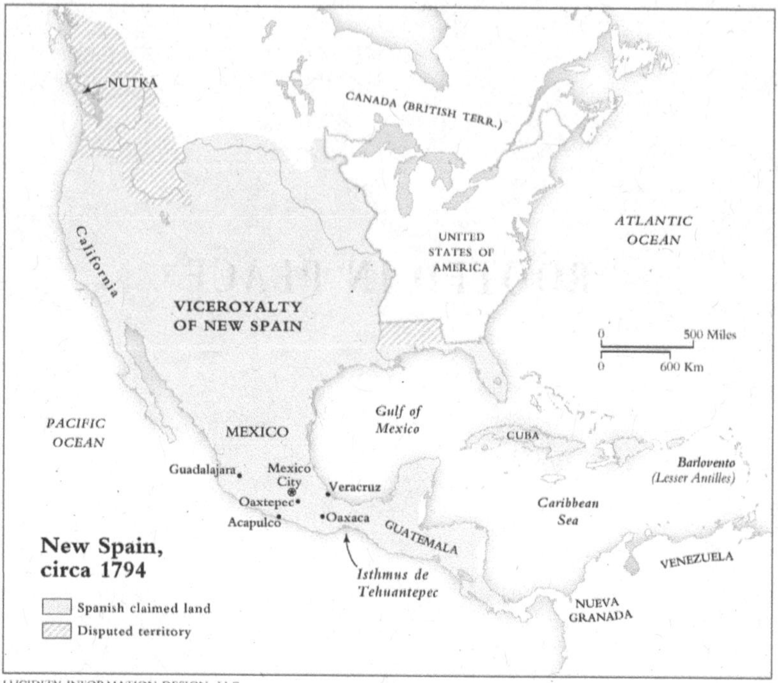

LUCIDITY INFORMATION DESIGN, LLC

MAP 1 New Spain, circa 1794.

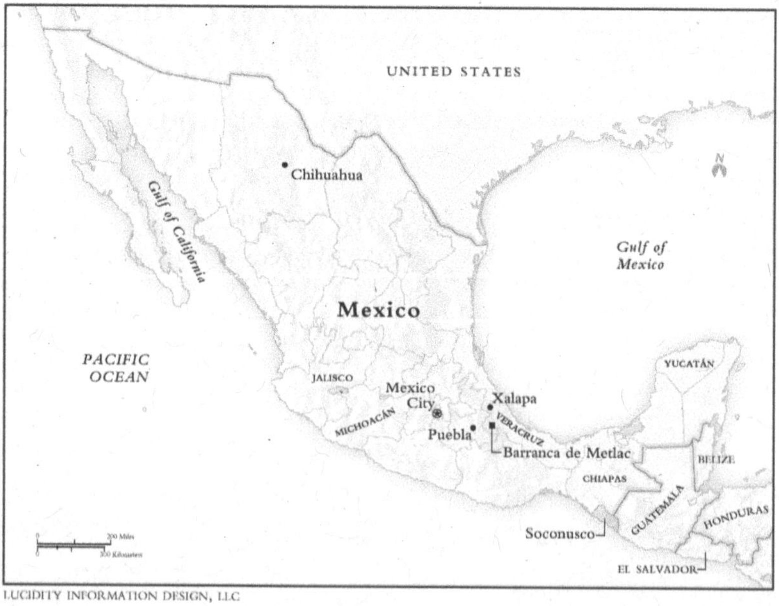

LUCIDITY INFORMATION DESIGN, LLC

MAP 2 Mexico.

INTRODUCTION

The Nature of Place

> He was educated on his father's estate by one of those Jesuit tutors who, though they were expelled by the king, managed to return in secular clothes to carry out their obsessive mission among us: to teach us that American flora and fauna exist, that there are American mountains and rivers, and, above all that we have a history that isn't Spanish but Argentine, Chilean, Mexican . . .
>
> —CARLOS FUENTES, *THE CAMPAIGN*

In 1893, a crowd gathered in the Mexican Pavilion of Chicago's Columbian Exposition to see the publication of recently rediscovered documents from the Royal Botanical Expedition to New Spain (1787–1803), including one of the most comprehensive lists of Mexican flora. Officials and scientists declared on this international stage that Mexico had at last joined the small club of modern, industrializing civilizations. And they presented the Royal Botanical Expedition's report as a proud assertion of their government's right to control its own natural resources.

It might seem strange that late nineteenth-century nationalists at this world's fair would choose a Spanish colonial institution as a symbol of Mexican national sovereignty. The choice comes across as all the more perplexing if we recall that the purpose of the Royal Botanical Expedition had been to assert European imperial domination over Mexico and its natural resources. To make sense of this seemingly odd choice, we need to go a bit further back in time . . . actually, a lot further back, to the 1500s. We need to return to the moment shortly after the Spanish conquest of the Aztecs, when the Spanish royal court physician Francisco Hernández arrived in Mexico City, sat down with native doctors known as *titicih* (sing. *ticitl*), and recorded what they told him about Mexico's plants.

Since the first moment of conquest, colonizers and the colonized alike confronted questions about what it meant to be from this place, what natural resources Mexico offered, and who had the right to control those resources and on what basis. Mexico is the third most biodiverse country in the world thanks both to its varied biogeography and to the long and distinctive history of human interactions with the region's flora and fauna before the arrival of Europeans.[1] This book is an account of how scientific intellectuals in Mexico laid claim to these diverse natural resources, first on behalf of the Spanish Empire and then in the name of the republic. It tells this story through three transformative and interlinked moments: (1) the royal expedition by Francisco Hernández during the late sixteenth century, which inspired naturalists to contemplate how Mexico's native plants and cultures fit into the expanding world of Renaissance knowledge; (2) the Royal Botanical Expedition of the late eighteenth century that set out to build on the earlier findings by Hernández, while imposing Enlightenment science as a tool for studying and conquering the botanical frontier in New Spain; and (3) the late nineteenth century, when leading scientific intellectuals looked nostalgically on previous expeditions as they sought to forge Mexico's future progress by harnessing nature, science, and indigeneity for the good of the republic.

These three moments, I argue, were links in a single chain in which scientific intellectuals (known as naturalists during most of the period covered by this book) debated what it meant to know and claim the flora rooted in Mexican soil—ranging from individual plants to forests and vegetated landscapes. And, in each link of this chain, they centered claims about the importance of indigeneity. The views of plants, place, and indigeneity that scientific intellectuals forged during these three crucial historical moments continue to shape current-day debates about what rights Mexico has to its natural patrimony, how these rights are distributed among its population, and how and why species and ecosystems should be protected.

NATURE AND THE IMAGINED COMMUNITY

Environmental historians sometimes argue that plants, animals, and ecological zones do not recognize national boundaries, and therefore neither should the scholars who study them. This argument, while well taken, can obscure the historical connections between national identity and the natural world. It also can distract us from the ways nationalist views of nature affect environmental policy. While it is true, for instance, that mountain lions wander freely between the U.S. and Mexico (at least they did until the United States built a steel wall along the border) and whales glide uncaringly from U.S. to Mexican waters, the fact is that the people and institutions of the U.S. and Mexico deal differently with these mountain lions and whales. The differing ways countries deal with conservation is connected to the ways the people of those countries conceive of the historical relationship between the natural world and their respective imagined national communities.[2]

The connections between identity and nature contributed to the earliest rumblings of nation formation. While integration of the masses occurred later than was sometimes presumed, it is clear that during the nineteenth century conversations about what defined the nation became widespread among the middle and upper classes throughout much of the world.[3] Intellectuals, politicians, and artists embraced their natural environment as one of the features that most anchored them to a particular place and distinguished them from other national communities. Historians of Germany, England, France, and the United States have been particularly attentive to this historical relationship, and their findings can productively inform the questions we ask about the relationship between imagined communities and perceptions of nature within Mexico.

According to historian Michael Imort, during the nineteenth century "German public discourses were replete with ethnic or *völkisch* interpretations that presented forest-mindedness not as a learned cultural pattern, but as a national characteristic of Germans that was supposedly the result of two thousand years of co-evolution between forest and people. In other words, forest-mindedness was presented as a defining racial element of Germandom."[4] Germans did not imagine their forests to be "wilderness," as in the United States (see below), but as part of a landscape saturated with cultural meanings that endowed their country with a unique identity that extended from the ancient past to the present and into the future.[5]

England developed its own distinctive relationship to nature, which in turn shaped the appearance of its countryside. In the 1500s and 1600s, the English idealized a regimented landscape, tamed into orderly wheat fields that surrounded expanding urban centers. Once most of the country had been conquered by this orderliness, people experienced nostalgia for a countryside that felt more natural. This led to a new aesthetic expressed in the sprawling gardens created by the aristocracy. This aesthetic ideal soon jumped the garden wall and transformed England's countryside. Despite being highly manipulated, the resulting landscape gave the illusion of naturalness. Appreciation of this ideal depended on at least some familiarity with landscape painting and popularized the term "picturesque" to describe a setting that had been manipulated to conform to the aesthetics of a picture.[6] By the mid-nineteenth century, these cultivated landscapes were incorporating Himalayan rhododendrons, African lilies, and Japanese maples, along with a growing number of other plants acclimated from across the globe, converting them into expressions of the power and overseas reach of the British Empire.

France experienced a similar evolution from cultivated gardens toward curated landscapes, but with a different outcome. Nineteenth-century France followed England in the move away from strict regimentation toward a naturalistic aesthetic. But instead of imitating the English landscape garden ideal, France, consistent with its motto *Liberté, Égalité, Fraternité,* moved toward an undulating rural peasant pastoral countryside, picturesquely divided into small farms or vineyards marked by hedges and dotted by quaint villages. Beyond this peasant pastoral setting lay forests rising into the Alps to the southeast and the Pyrenees to the southwest.[7] Both the English and the French landscape were intentional constructions based on how the people of each country understood themselves and their relationship to their natural world.

In the United States, writers, artists, and foresters drew inspiration from German and English Romanticism but remade them to fit a new context. The connection between nature and nation in the United States found expression in painters' and poets' quasi-religious devotion to wilderness, whether in the Hudson Valley or Yosemite. Historian Roderick Frazier Nash goes so far as to argue that the idea of "wilderness was the basic ingredient of American [i.e., the United States] culture," giving its "civilization identity and meaning." They imagined this wilderness as nature's cathedral, channeling God's favor on the United States just as the cathedrals of Notre Dame or Chartres did for France. By the end of the century, Frederick Jackson Turner (1861–1932) famously drew on this vision of the wilderness frontier to make a case for American exceptionalism.[8]

Nineteenth-century Mexicans were part of this same trend, sometimes in conversation with, but never simply imitating, their U.S. and European contemporaries. Notably, Mexicans pondering the meaning of their natural world rarely mentioned Spain. Much of the explanation comes from the fact that Spain was a latecomer to such discussions. Its artists and intellectuals did not take up such questions until after the crisis of 1898, when Madrid lost the last of its overseas empire after the Spanish-American War.[9]

Environmental conservation policies tend to be most effective when they enjoy broad support, which is gained by tapping into the ways national identity and environment are intertwined within public discourse. This is clear, for example, in the United States, where the wilderness ideal endowed the national identity with a natural heritage that the public deems worthy of protection. Assumptions about wilderness have brought their own problems, but this does not negate the fact that the entanglements between U.S. national identity and wilderness made it easier for conservation policies to resonate with the public. In the case of Costa Rica, often heralded as Latin America's environmental success story, the existing literature and national attitudes might leave one with the impression that the ecology of Costa Rica stops at the southern border with Panama and the northern boundary with Nicaragua. This belief in the uniqueness of the U.S. environment as a space for individualism and democracy and of Costa Rica as the "Green Republic" speaks to the ways in which views of the environment are bound tightly to discourses of national identity in ways that can support conservation.[10] While Costa Rica and the United States, along with France, Germany, England, and even Australia, New Zealand, Canada, South Africa, Japan, and many other countries, all exhibit compelling nationalist ecological imaginations that contribute to their management of nature within their domestic territories, it is clear that they do not all share the *same* nationalist ecological imagination.[11] The linkages between nation and nature in the U.S. differ from those in Germany, which differ from those in Costa Rica or Japan. Each is rooted in the particular political, cultural, and environmental history within the territory claimed by their respective state.

In the United States and some other countries, the history of these relations has been well studied, but this is not the case within most of Latin America, where we lack the historical guidance necessary to craft conservation policies that resonate with the history and collective self-perception in each country. This book helps fill this gap by tracing the connection between nature and identity formation within Mexico and perhaps might point the way for similar studies of additional countries in Latin America and other parts of the world.

LATIN AMERICAN ENVIRONMENTAL HISTORY

When talking about the environment, it is easy to get lost in declensionist narratives of gloom and doom. There is good reason for this, considering the rapid rate at which humans are transforming our natural world. But, as environmental historians Mark Carey and Claudia Leal remind us, there are other narratives about our relationship to the natural world that need to be told.[12] This is one of those narratives. It is in part a history of science, recounting how naturalists during the Renaissance, the Enlightenment, and then the nineteenth century went about studying the natural world in Mexico. It is also a cultural history, reconstructing how people understood themselves and one another, and their relationship to the plants and landscapes around them. And it is a political history, tracing how colonial structures of rule were created, how indigenous populations were subjugated, and how a republic emerged from the ashes of colonialism. It also is art history, tracing what it meant to see and depict nature, and how visual aesthetics were crucial to assertions of empire and nationhood. It is all these things. But the thread that binds them together is environmental history, understood as the history of human interactions with the natural world.

More specifically, this study is *Latin American environmental history*. Whereas U.S. environmental history arguably was born out of a concern for wilderness conservation and only gradually discovered the importance of social justice issues, for most scholars working on Latin American history, interest in the environment sprang from efforts to understand social inequality and power relations, including imperialism, class conflict, racism, and struggles over resources.[13] By approaching questions through the lens of environmental history, we can account for the connections between society and ecology, between science and culture, and between understandings of nature and political structures of domination, recognizing these connections as drivers of historical change.

Despite their shared interest in social justice, there exists no uniform or singular approach among practitioners of Latin American environmental history. Studies within this field tend to be interdisciplinary and to cut across thematic approaches. Nonetheless, to contextualize the approach of this book, it is useful to organize existing contributions into general thematic and methodological approaches.

First, there are a number of broad overviews as well as case studies that helped define the field and set an agenda for future research.[14] There also are various studies of agricultural commodities and of natural disasters, and a rich literature on forestry, deforestation, conservation, and park creation.[15] Some environmental historians have examined frontier zones, while others have focused on urban areas and adjacent aquatic spaces.[16] And a large body of scholarship has brought innovative insights into our understanding of conflicts over agricultural land, water resources, mining, and fossil fuel extraction.[17] Scholars of historical geography have informed our understanding of how space and landscapes overlap with ecological change and political consolidation.[18] Some works even analyze art and aesthetics in relation to environmental history, though this remains a more sparsely developed area of scholarship.[19]

There also are related fields in which practitioners might not necessarily take an overtly environmental historical approach. This is notable in the large and impressive literature on the history of science in Latin America, especially of botany. Most of these works steer surprisingly clear of environmental questions, but others have placed their work intentionally at the intersection with environmental history.[20] Debates over the fate of the commons also has played an important role in defining pressing research questions.[21] And, more recently, studies that import a particularly U.S. approach toward the study of environmental justice have developed into a growing field, with a number of its researchers giving at least some attention to historical questions.[22]

This book, like many of those listed above, strategically interweaves different approaches. It draws on histories of science, landscape studies, visual culture, race and ethnicity, empire, and nation formation to trace the ways scientific intellectuals within Mexico viewed, described, visually depicted, and attempted to alter their natural world, and how their efforts contributed to structures of domination and political organization. While this book is not a history of conservation, it addresses past efforts to manage the natural world, and the conclusions it draws regarding the relationship between nature and social-political constructions of empire and nation have implications for how we formulate effective conservation policies in the future.

IMAGES, ARGUMENTS, AND ORGANIZATION

This study is based mostly on archival and printed sources produced from the sixteenth century through the end of the nineteenth. These include correspondence, government reports, scientific studies, maps, and diagrams, as well as books and journals published at the time. Special mention needs to be made of the use of visual images such as scientific illustrations, landscape art, and even charts and tables.

There is nothing straightforward about a visual depiction, just as there is nothing straightforward about a written document. The question of how to integrate objects and images into their studies remains a bedeviling one for historians, who frequently subject them only to haphazard analysis or else use images as ornamental supplements to the text. Images can serve as historical documents in their own right, but they often require different analytical strategies than those we use when analyzing written sources. How then, can we productively "read" them? Each of the three sections of this book takes up this methodological challenge. Renaissance and Enlightenment botany were profoundly visual enterprises that operated through the creation of gardens, herbaria, and scientific illustrations. I employ the term "visualization" in recognition of the degree to which officials, botanists, and draftsmen not only produced imagery, but also disciplined how they and others viewed nature. This book, therefore, subjects visual images to the same level of critical analysis as it does documents, while recognizing the ways that images need to be approached differently than texts.

The use of visual and material culture in conjunction with other types of historical documents also can offer glimpses into the life, work, and contributions of a range of people from the past, including the sorts of individuals whose stories rarely appear in historical accounts but whose decisions and actions are as important as those of people who usually are centered by historians. The contributions by people such as gardeners, construction workers, laborers, and illustrators, who physically created the material world around us, are usually left out of historical accounts in favor of people such as patrons, architects, landowners, and politicians who bequeathed an abundant paper trail. As these chapters show, bringing in these lesser-known figures does not just round out our appreciation of the past, but rather changes our understanding of the processes of historical transformation.

Part 1 of the book focuses on the Francisco Hernández expedition (1570–1577) undertaken by the Spanish king's *protomédico* (chief medical officer) of the Indies, which was the first scientific study of flora in the New World. The chapters in this section address the interactions between Francisco Hernández and the native Nahua peoples. Through their strategies for categorizing and visually depicting nature, Hernández and his Nahua collaborators created a distinctive and enduring understanding of indigeneity, nature, and territoriality that later would serve as a foundation for the construction of a form of nationalism rooted in the region's environment. Within this first part of the book, chapter 1 argues that, despite clashes between indigenous society and European colonizers, there existed opportunities for a surprising level of cross-cultural collaboration, as well as intellectual improvisation. Chapter 2 argues that Hernández did not simply transmit the information provided by his native informants. Instead, he adapted their system of organizing knowledge, as expressed through their taxonomic nomenclature based on Náhuatl, for use in Europe and, eventually, he hoped, around the world. His fascination with the potential for Nahua methods to serve as part of a universal system extended into native approaches to visually depicting nature. The chapter shows how the techniques of botanical illustration devised by the native draftsmen hired by Hernández challenged emerging European notions of what it meant to accurately depict the natural world. Hernández's taxonomic and representational proposals received a cool reception in Europe, where naturalists tried to restructure his texts to fit existing European theories and purged his manuscripts of the images produced by native draftsmen. Nonetheless, his efforts inspired scientists in the coming centuries, and even into our own day. Taken together, these two chapters argue that the Hernández expedition placed indigeneity at the center of emerging understandings of Mexican nature, and that this would have an enduring impact on how nature and empire (and, later, nation) were constructed in relation to one another.

Part 2 focuses on the Royal Botanical Expedition to New Spain (1787–1803). The original purpose of this Enlightenment-era expedition was to reassert Spanish imperial domination over New World resources by updating and modernizing the Hernández expedition from two centuries earlier. While the expedition's members continued to emphasize native plants, they took a dim view of indigenous people and their knowledge. The expedition also ended up going far beyond the ambitions of its sixteenth-century predecessor in terms of exploration of the botanical frontier and in its effort to reshape local practices and institutions within Mexico. Chapter 3, the first of this section's three chapters, traces the origins and activities of the expedition, as well as its importance to Enlightenment science and local governance. It pays particular attention to the Royal Botanical Garden and the school of botany and medicine that the expedition created in Mexico City, since these were branches of the expedition designed to most affect local Mexican society. The chapter argues that, in contrast to Hernández's respect for native knowledge, members of the Royal Botanical Expedition imposed a vision in which expertise regarding the natural world was to be hierarchically ordered, dominated by Enlightenment men of science, and disciplined according to a single structure that relegated alternative modes of knowledge or organization, especially those derived from indigenous people, to the private sphere as quaint, ethnically specific, feminine, and folkloric. Moreover, the expedition's garden and school engaged in two levels of mimicry that shaped views of Mexican nature: (1) imposition of scientific institutions modeled after those of Europe; and (2) creation in Mexico City of a botanical microcosm of the plants found across New Spain. Even as the expedition subordinated the colony to the metropole and overtly belittled indigenous knowledge, it also nurtured possibilities for conceiving of New Spain as a distinct imagined community rooted in nature, centered in Mexico City, and dependent on the forms of indigenous knowledge it overtly disparaged.

Chapter 4 examines how members of the Royal Botanical Expedition used the innovations introduced by Swedish botanist Carl Linnaeus to figure out which of the plants they encountered in Mexico were new to modern science, what the most important characteristics were of these plants, and how to fit them into emerging structures of knowledge. Much like the colonial process described by historian Elinore Melville, which transformed the land into something alien to the native people who had been most intimately familiar with it, so the scientific practices imposed by the Royal Botanical Expedition, including what I term its "Linnaean optic," created structures of knowledge that separated native people from knowledge about the natural resources with which they had been most familiar.[23] Chapter 5 traces the fate of the Royal Botanical Expedition and its findings, both in Europe, where it ended up meeting an end eerily similar to that of its sixteenth-century predecessor, and in Mexico, where its institutions and intellectual structures became the foundations for a national science after the consummation of Mexican independence in 1821.

Part 3 of the book argues that scientific intellectuals of the second half of the nineteenth century engaged in a project of nostalgic recovery on behalf of the Mexican Republic, in which they sought to make up for earlier lost opportunities—foremost among which was the failure to exploit the findings of the Hernández expedition and the Royal Botanical Expedition. It argues that, in contrast to the United States, where nationalist interest in nature centered on the idealization of wilderness, in Mexico scientific intellectuals, similarly to their German counterparts, emphasized nature's association with history. Mexicans particularly stressed their nature's rootedness in indigeneity, even as elites seized control of the land and resources from living indigenous people. Chapter 6, the first chapter in this section, traces the process of nostalgic recovery by patriotic scientific intellectuals. It shows how and why they looked to particular native plants and aspects of indigenous heritage, and how they embraced colonial expeditions to construct a specifically Mexican scientific tradition. As described by Eleuterio González in 1882: "Hernández," by documenting Aztec knowledge, "lay the foundations for Mexican Natural History, Cervantes [of the Royal Botanical Expedition] and his many disciples cultivated it with diligence; and the Sociedad Mexicana de Historia Natural brought it to the level found in the most learned nations of Europe."[24]

Nineteenth-century scientific intellectuals were particularly eager to demonstrate how their individual efforts added up to something larger that could serve the needs of the republic and its citizens. Chapter 7 argues that they became convinced that they could best achieve this goal by setting aside the dispassionate abstraction idealized by their Enlightenment-era predecessors in favor of an approach that emphasized the Romantic aspects of their explorations, including their own role as heroic eyewitnesses to discovery in the field and in the laboratory. Their heroic narratives rooted Mexican nationhood in the distinctiveness of its plants, forests, and landscapes, and the capacity of leaders such as themselves to exploit these natural resources for the good of the republic. This desire to bridge science and Romance gave rise to a novel way of picturing Mexico's natural landscape, best expressed through the art of the scientifically trained landscape painter José María Velasco, which in turn inspired new ways of using visuality in support of nationalist science. It also inspired medical professionals to build on Mexican plants and culture to cultivate a distinctively Mexican approach to modern medicine.

As they defined the natural world around them as national patrimony that could contribute to the economic prosperity of the republic and the well-being of its people, scientific intellectuals became concerned about the rampant rate at which this inheritance was being squandered, above all through deforestation. This is the topic of chapter 8, which argues that Mexicans scrambled to find a way to describe what was being destroyed and why it mattered, to figure out who was to blame, and to identify effective policies for conservation.

The book closes with a conclusion and a postscript that show the surprising ways the transformations described in the previous chapters continue to shape Mexican identity and environmental policy, from the design of its national flag to the creation of national seed banks, and even how they shape international approaches toward conservation. Together, the conclusion and postscript propose future pathways through which policymakers might draw on this history to better protect the nation's biological diversity and conserve its natural resources for the good of the Mexican people.

PART I

THE FRANCISCO HERNÁNDEZ EXPEDITION, 1570–1577

CHAPTER 1

RENAISSANCE BOTANY IN MEXICO

The Francisco Hernández expedition (1570–1577) was the first European scientific study of New World nature. It was undertaken at the height of Spain's golden age by one of the era's best-trained naturalists, as scientists were known at that time. During his medical training at the Universidad de Alcalá de Henares, Hernández had mastered the classic works by Aristotle, Theophrastus, Pliny the Elder, Pedanius Dioscorides, Hippocrates, Galen of Pergamon, Avicenna (Ibn Sina), and Mondino de Luzzi. After university, he achieved status within Spain as a researcher and a practicing physician, leading to his court appointment in January 1570 as *protomédico* (chief medical officer) of the Indies and head of the King's natural history expedition to New Spain.[1] Upon his 1571 arrival in Mexico City, Hernández found that the Aztec gardens, doctors, herbalists, and illustrators were still alive and well, and he quickly began collaborating with *titicih* (native doctors) and *tlacuiloque* (draftsmen) to bring Nahua and Spanish intellectual traditions into direct dialogue. The findings of his expedition contributed to a revolution in Renaissance natural history and laid the foundations for Mexican botany.[2]

Most scholarship on the Hernández expedition focuses on its contributions to European botanical science or uses his findings to trace the history of particular plant species. The chapters in this section, by contrast, turn attention to the interactions between Francisco Hernández and the Nahua to shed light on the emerging connection between nature and empire. Through their strategies for categorizing and visually depicting plants, Hernández and his Nahua collaborators created an enduring understanding of indigeneity, nature, and territoriality that later inspired a follow-up expedition during the Enlightenment. I argue that, over the long term, it would also come to serve as a foundation for the construction of Mexican nationalism rooted in nature and indigeneity. The first thing to figure out, however, is what kinds of connections Hernández forged with his native collaborators.

NEW WORLD ENCOUNTERS: COLLISION OR IMPROVISATION?

Hernández's findings astonished European naturalists by introducing them to more than three thousand New World plants. To appreciate the enormousness of his contribution we should recall that, until Hernández, many European scientists still believed that the ancient inventories compiled by Theophrastus (c. 371–c. 287 BC) and Dioscorides (c. 40–c. 90 AD), which listed between three hundred fifty and five hundred species, were comprehensive lists of all the world's plants.[3] Hernández's twenty-four books and his ten volumes of botanical illustrations, all of which were created by indigenous draftsmen, increased six-fold the number of plants known to science, forcing Europeans to reconsider what they thought they knew about God's creation, and leaving them dumbfounded about how to assimilate such a sudden, massive contribution.

Earlier accounts of the New World had been written by chroniclers with no scientific training. Notable examples include Christopher Columbus's letters describing his 1492 encounter with the people of the Caribbean, Bartolomé de las Casas's bone-chilling description of atrocities committed against the native population, and the famous accounts of the 1519–1521 military conquest of the Aztec empire written by Hernán Cortés and Bernal Díaz del Castillo. One of the first accounts before Hernández that included a careful description of nature was *De la natural historia de las Indias* (*Of the Natural History of the Indies*), published in 1526 by Gonzalo Fernández de Oviedo y Valdés (1448–1557). Rather than attempt scientific analysis of the natural world, Oviedo took the point of view of an eyewitness reporting on the wonders of the Caribbean. Whether to attract readers or as part of his own intellectual effort to reconcile the new with the old, he combined direct observation with perpetuation of stories about wondrous creatures long thought to exist beyond the known world, such as the Amazonian women of Greek mythology.[4] When Hernández visited Mexico from 1571 through 1577, his was the first concerted effort to scientifically study New World nature, separating fact from fable and systematically reconciling it with existing knowledge in Europe.

Too often, scholars' narratives about the history of science reduce it to teleological accounts of how particular individuals either helped scientific fields advance toward their modern-day configurations or else stood in the way of such progress. The history of science within Latin America tends to maintain this basic framework while asserting that scientists in places like Mexico or Argentina made greater contributions to the development of particular scientific fields than has generally been acknowledged. Critics of the history of science, meantime, too often stereotype Western science as a monolithic tool of oppression while romanticizing other forms of knowledge as counter-hegemonic acts of resistance against European domination. Such approaches risk reducing non-Western actors as either complicit or resistant, with little acknowledgement of the complexity and creativity of human interactions and agency.[5]

We would be wise to eschew the preconceptions too common in these approaches by focusing, instead, on the processes of improvisation and negotiation as each of the historical actors attempted to communicate with one another while navigating a changing world. By considering how people at the time saw, described, and depicted plants and other parts of nature, with particular attention to their systems of nomenclature and their approaches to botanical illustration, the chapters of this section center these historical actors' humanity and agency. They trace how Hernández and his Nahua collaborators improvised in their use of linguistic and visual practices to interpret Mexican nature and to define its relationship to indigeneity and universality. This was not a simple two-sided encounter of European versus native. Instead, the encounter took place through the lives of individuals struggling to learn from one another even as they navigated demographic crises and rapidly changing configurations of power. These individuals were eager to bridge worldviews that were not easily translatable and demonstrated remarkable intellectual flexibility and ingenuity as they dealt with the challenges.[6]

In her studies of science and empire, Londa Schiebinger has analyzed the entanglements between political imperialism and scientific practice. She draws attention to what she terms "agnotology," by which she means culturally induced ignorance within the practice of science. Unable or unwilling to reconcile some of the new information with their own firmly held assumptions, Europeans dismissed certain kinds of information as "noise" rather than recognize it as intellectually coherent.[7] While such encounters were the norm, as demonstrated by the cool reception the protomédico's findings later received in Europe (see chapter 2), examination of Hernández's interactions with Nahua titicih offers novel insight into a moment when fertile cultural exchange still seemed possible. Though European scientists ultimately rejected Hernández's experiments in transculturation, his efforts influenced how later naturalists would think about the relationship between indigeneity and scientific understandings of Mexican nature.

"A CHAOS OF NOMENCLATURE": NEW TECHNOLOGIES IN THE STUDY OF PLANTS DURING THE RENAISSANCE

Thanks to his university training, Hernández was well aware of the difficulties involved in naming plants and the problems this caused for research and medicine. During the Renaissance, most medicines came from plants and were known as "simples." Naturalists who studied these simples sometimes fixated on the words of the ancients, guided by an assumption that close analysis of the original texts of Dioscorides, Galen, Theophrastus, and Pliny was more important than anything that they might gain by looking at a flower plucked in a meadow. When they did directly observe nature, it was often to refine their reading of the ancient texts. As the French Enlightenment philosopher Jean Jacques Rousseau later mocked in his 1785 botany manual: "at the revival of learning, everything [i.e., localized vernacular understandings of plants and their uses] disappeared to make room for the works of antiquity; nothing was then either good or true but what was to be found in Aristotle or Galen. Instead of searching for plants where they grew, they studied them only in Pliny and Dioscorides, and there is nothing so frequent in the authors of those times, as to find them denying the existence of a plant, for no other reason, but because Dioscorides did not mention it in his Materia medica."[8] Though hyperbolic, there is truth to Rousseau's assessment.

Yet things were beginning to change. Renaissance naturalists were discovering that the ancient texts were less comprehensive than had been presumed. And so they began to see themselves not just as recreators of ancient knowledge, but as producers of new knowledge.[9] This change was starting to take hold when Hernández completed his studies at the Universidad de Alcalá de Henares and accelerated greatly after his return from Mexico in 1577.

Even as they began to realize the limitations of the knowledge passed down by the ancients, Renaissance naturalists clung to their classical texts. They did so for at least two reasons. First, just as Latin gave intellectuals across Europe a shared language for communication, so too did these ancient texts, despite their ever more apparent limitations, provide naturalists with shared reference points for the accumulation of scientific knowledge about the natural world. Second, as scientists encountered a growing number of plants, they needed a way to distinguish what was novel from what already was known. Some of the more cutting-edge naturalists, including Hernández, did this by using the emerging field of philology to compare texts and translations in an effort to engage with the ancient texts in their original languages and systems of conceptualization, and thereby discern as best they could what, precisely, the ancient authors had originally described. The hope was that, once they had deciphered precisely which plants Pliny and Dioscorides had described, they could more confidently add their own discoveries to this inventory and eventually devise a master list of all God's creation.

The effort to systematize what was known to add new information demanded networks of knowledge and long-distance communication among scholars.[10] It also required new technologies such as scientific botanical gardens in which to grow and observe species of interest, and herbaria in which to archive dried plant specimens for future reference—both of which were introduced in the 1540s in Pisa.[11] Gardens and herbaria helped researchers working in situ, but they did little to aid long-distance communication amongst scholars. Long-distance communication demanded something different, a third new strategy: botanical illustration.

Scholars communicated with one another by writing letters and publishing texts. When a naturalist wrote a colleague about a particular plant, they (or "he," since it almost invariably was a male) needed a way to assure that his reader had the same plant in mind. This was more complicated than it might seem because scholars lacked a standard system of nomenclature; that is, of naming things. A single plant often went by dozens of different vernacular names and perhaps another dozen Latin names. As Rousseau explained: "... since translators, commentators and practitioners seldom agreed in their choice, twenty names were given to the same plant," and "every man maintained that his own was the true one." There "was such a chaos of nomenclature, that the Physicians and Herborists [*sic*] no longer understood one another: there was no possibility of communicating their mutual lights, nothing remained but disputes of words and names, and even every useful enquiry and description was lost, for want of being able to decide what plant each author had spoken of."[12]

Naturalists decided that inclusion of a drawing of the plant to which they referred could increase the likelihood that they and their readers might have in mind the same plant. But the arbitrary and decorative manner in which medieval European and Muslim draftsmen had illustrated plants was of little use. And so Renaissance botanists and artists experimented with new strategies of visualization.[13]

The attraction of visualization, rather than only written description, made sense in an era when Renaissance artists were creating greater faith in the descriptive ability of visual images by incorporating mathematically accurate depictions of three-dimensional spaces into their drawings and paintings. In 1530, Otto Brunfels (1488–1534) published *Herbarum vivae eicones* (*Pictures of Living Plants*), the first widely read European herbal that included images. At a time when most plant illustrations were either fanciful depiction or else simply copied from previous woodcuts, which themselves were copies of other previous works, the images that the artist Hans Weiditz (1495–1537) created for Brunfels's herbal were drawn directly from plant specimens. In doing so, Weiditz followed the instructions of his mentor, Albrecht Dürer (1471–1528), the prominent German artist who advised plant illustrators "Do not depart from" nature, "for art is hidden in nature and he who can draw it out possesses it."[14]

The problem for botany was that there was no consensus about what it meant to accurately depict a specimen. The naturalist Vicente Danti

(1530–1576) differentiated between drawing "things as they are seen" versus depicting them "perfectly as they ought to be seen." It was to the second that the naturalist should aspire.[15] But this left unresolved the problem that a draftsman depicting a particular specimen had no way to know which aspects of its appearance were typical of that kind of plant (hence, how it "ought to be seen"), and which were accidents of its growth (e.g., latitude and elevation) or the result of funguses, insects, or weather.

The difficulty of determining which aspects of a plant's appearance were characteristic was heightened by the lack of consensus about what parts of the plant to emphasize in written or visual descriptions. Many draftsmen hedged their bets by trying to emphasize everything. Others, sensing the impossibility of producing a totalizing image, arbitrarily zeroed in on one or two features, such as a fruit, root, or leaf. A good example of botanical draftsmanship of the era is the *Neuw Kreuterbuch* (*New Herbal*) that Jakob Dietrich (aka Jacobus Theodorus Tabernaemontanus, c. 1522–1590), often included as one of the so-called "Fathers of German Botany," assembled during the years that Hernández was in Mexico and published shortly after the protomédico's return. To maximize the odds that he and his reader might be talking about the same plant, Dietrich provided numerous names for each plant, in as many languages as possible, including German, Latin, Greek, Spanish, Arabic, English, Italian, Dutch, French, Flemish, and Bohemian (Czech). To further improve the odds of effective communication, he included illustrations of every plant, each of which was designed to convey the impression of verisimilitude between itself and the not-present specimen.[16]

In his early sixteenth-century account of nature in the New World, Fernández de Oviedo had incorporated schematized depictions of leaves, fruit, or sometimes whole plants within his narrative. Oviedo's thinking about accuracy was more connected to his impressions of the art he had seen in Italy than to the emerging practices of botanical illustration that preoccupied Hernández and Dietrich. And Oviedo's images were geared toward bolstering his status as eyewitness rather than to advance the reader's understanding of plant science. In his description of a tree, for example, Oviedo stated, "I will relate what I have learned about it [the tree], deferring to whomever knows best how to paint it or make it understood, because it would be preferable to have it rendered by the hand of . . . painters as excellent as . . . Leonardo da Vinci, or Andrea Mantegna, famous painters I got to know in Italy, than to describe it by means of words. All of this [New World nature] is better seen than written about or painted."[17] Images of pineapple and other plants that Oviedo included in his book were designed to help the reader join the author in bearing witness to New World wonders, not to assist in scientific classification.

Dietrich and Brunfels, by contrast, were not concerned with elevating their status as eyewitnesses, but in trying to use representation to understand a plant scientifically, and to share that knowledge over distance with other experts. Their images decontextualized each plant from its place of origin and its environment by placing it on a blank background, surrounded only by Latin text. Such images represented a major advance compared to vague medieval plant illustrations, and an alternative to the eyewitness approach taken by Oviedo. But even when paired with the long list of names, they left too much uncertainty between the author and his distant readers scattered across Europe and beyond.

The illustration of coriander, or cilantro, on his page 217 is typical of Dietrich's approach (see figure 1). It is a stunning image, with a level of detail that was impressive for its time. Almost two centuries later, during the Enlightenment, the Swedish naturalist Carl Linnaeus (1707–1778) would teach draftsmen to zero in on particular features of a plant—primarily its reproductive organs—with exaggerated clarity. Dietrich's draftsmen, however, gave all parts of the plant equal attention: the roots, the stem, the leaves, and the flowers. And despite close attention to numerous visual details, these are rendered in a stylized manner. Confusions caused by this stylization were heightened by the fact that all plants were shown the same size regardless of their relative differences in height, and all are noticeably cramped to fit within the rectangular dimensions of the printing block, thus distorting our perception of the plant's growth habit. Historian of science Brian Ogilvie has shown that "the crudeness of botanical illustration" of the era "was not due [just] to technical limitations in the medium [woodblock] but rather to artistic intentions."[18] Contortion and stylization in Dietrich's images extend to the lack of attention to the relative differences in size between flowers and leaves in most of the plants, and also regarding the length of leaves or basal stems as compared to flower stalks.

In addition, some of Dietrich's plants are depicted during their early spring growth, others while flowering, and others while bearing fruit or going to seed. We can see some of the implications of this arbitrariness in

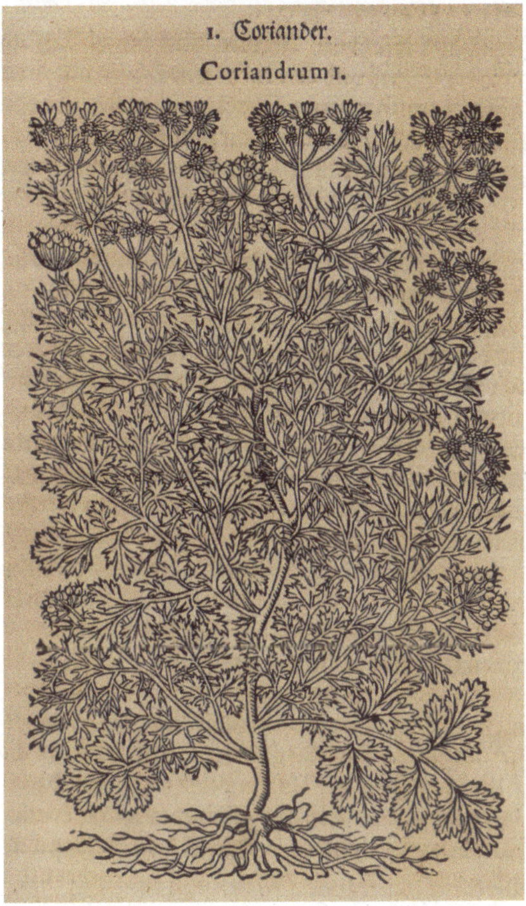

FIGURE 1 Coriandrum, from Jakob Dietrich, *Neuw Kreuterbuch*, 217. Courtesy of Amherst College Archives and Special Collections.

his image of maize (corn), which had recently been introduced from the New World. It depicts only the ears, which appear partially peeled so that the artist could show in great detail the accidental patterns of the teeth. Such images give us a good idea of what corn looked like, but they would have been of little to no use to someone trying to figure out what variety of corn he was holding in his hand or how it might be related to other plants.

While Hernández may not have known the work of Dietrich, he was keenly aware of the problems of nomenclature and illustration that Dietrich was trying to overcome, as well as the limitation of the kinds of solutions presented by Dietrich. He also would have recognized the limited scientific use of the style of representation employed by Oviedo.

The need for solutions became urgent for Hernández because he needed to figure out which of the plants he encountered in New Spain were new to European science and which were not. Fortunately, during his years in university at Alcalá, Hernández had been surrounded by colleagues who had labored on the famous *Complutensian Polyglot Bible,* in which they employed the latest techniques of philology to compare and reconcile Greek, Latin, and Hebrew biblical texts.[19] Hernández would bring this philological approach to bear on his experiences in Mexico as he sought to get as close as possible to the original native sources of information, and to think critically about what was involved in the translation of meaning between languages and across worldviews. In essence, Hernández extended cutting-edge philological techniques from religion into the burgeoning fields of natural history and cross-cultural communication with the hope of finally devising a solution for the "chaos of nomenclature."

"SEARCHING FOR THE SECRETS OF NATURE IN DISTANT REGIONS": A RENAISSANCE BOTANIST IN MEXICO

After brief stopovers in the Canary Islands, Haiti, and then Cuba, Hernández presented his credentials to a High Court representative at the Mexican port in Veracruz, then hurried to Mexico City. As soon as he established himself in the capital, he began meeting with native titicih to learn about medicinal plants, and he hired tlacuiloque to create botanical illustrations of the plants he studied. A series of letters between Hernández and the king during the expedition reveal how the protomédico attempted to carry out his charge. Scholars often use these letters and other texts by Hernández to recreate aspects of preconquest Nahua culture or to untangle plant species. Instead, I want to use these documents to gain insight into the process of improvisation by Hernández and by his Nahua collaborators, at a moment when their respective cultures were undergoing accelerated flux through their contact.

Hernández's actions in New Spain were grounded in how he understood the charge he was given by the king. Philip I explained that he sent Hernández to New Spain, rather than to one of his other colonies, "because we are informed that more plants, herbs, and medicinal seeds are to be found there than elsewhere." Rather than search for plants on his own,

Hernández was to rely on "doctors, medicine men, herbalists," and other native "persons with knowledge in such matters" in central Mexico as well as the provinces, writing down what they told him and organizing this information for use by European naturalists and royal officials. Far from looking down on or dismissing indigenous knowledge, the Crown signaled a surprising level of openness. Hernández's training in philology at Alcalá prepared him for the complications involved in translating across different languages and worldviews, and the king's orders created a mandate for him to put these skills and this open-mindedness into action.[20]

The king entrusted Hernández with a remarkable amount of leeway. Though the royal instructions made clear that the main task was to describe plants that had medicinal potential, the Crown encouraged Hernández to take a broad view of his charge by gathering "information generally about herbs, trees, and medicinal plants." In addition to finding out the medical properties of plants and how they were applied, he should ask the natives about "the places in which they grow and their manner of cultivation, and whether their habitat be dry or moist, or if they grow among other trees and plants, and if they occur in different varieties," and whatever other descriptions the natives might offer.[21] In short, Hernández was to learn the cultural and medical significance of each plant, along with its environmental and ecological context.

Philip II ordered Hernández to separate fact from fable by confirming all his information through direct experience, and then to vouch for the veracity of all that he recorded from the natives.[22] Hernández interpreted this to mean that, by testing and synthesizing what the natives told him, he also was to make his own original contributions to natural history, eventually describing his task in New Spain as "searching for the secrets of nature in distant regions."[23]

The more Hernández learned from his native collaborators, the more aware he became of how difficult it would be to assimilate what he was learning into the Renaissance structures of knowledge that he had mastered in Spain. He found that it was impossible to simply pluck information out of one system and insert it into the other. And he became convinced that he would have to draw on both European and native systems to devise approaches to nomenclature and botanical illustration that might overcome the limitations of European science, and thereby make it possible to absorb the vast new body of knowledge he encountered in

Mexico. In the end, his effort to comply with the king's instructions sometimes led him to demonstrate a level of respect for indigenous knowledge that would go far beyond what local elites, and even Crown officials, found acceptable.

LEARNING FROM THE TITICIH

Earlier European explorers in the Caribbean, such as Oviedo, had considered natives part of nature. Even when they relied on information from the local people, they tended to treat this knowledge as innate, rather than as acquired through intellectual rigor or structured inquiry. But in Mexico, where the Spanish colonists encountered Mexica civilization with its highly developed systems for exploiting and managing nature, they could sustain no such conceit. They recognized that they would have to rely on native expertise for harnessing the medicinal, nutritional, and commercial potential of New World plants.[24] Efforts to access native knowledge were aided by the activities of the Franciscans and other missionaries, who investigated every facet of native languages and culture that they thought might aid them in their proselytization, including natives' botanical knowledge and views of nature. Hernández was able to draw on the work by these missionaries and their native proselytes.

Until the arrival of Europeans, the native physicians known as titicih received formal training in an Aztec school known as a *calmecac*. When the Spanish conquerors dismantled the Aztec schools, the next generation of titicih received its education at the Real Colegio de Santa Cruz de Tlaltelolco (which I will discuss later) and through apprenticeships. The smallpox epidemic of 1519–1520 and the 1545 outbreak of a disease known as *cocoliztli* (literally, the pest) decimated the ranks of these native doctors. Despite this loss, when Hernández arrived in New Spain, he encountered enough titicih to guide his research, even if their knowledge and training had become fragmented compared to that of their predecessors. He also consulted the lesser-trained herbalists who served the popular classes, along with native apothecaries, *boticarios*. The earliest report about native apothecaries came from Hernán Cortés in 1520, shortly after his entry into the Aztec capital of Tenochtitlán. He described neighborhoods filled with shops in which "all the medicinal herbs and roots found in the land are sold," along with "ready-made medicines . . . ointments and plasters."[25] Hernández's contemporary, Francisco Cervantes de Salazar (1521–1575), described native boticarios who sold "ointments, syrups, waters, and other things for the sick. Almost all illnesses are cured with herbs, such that they even have a widely known herb to kill lice."[26] Beyond these formal apothecary shops, the streets and plazas of the capital were filled with street vendors hawking every kind of medicinal herb. The Mexican world that Hernández encountered in 1571, then, was one in which titicih and native boticarios had declined in number and training but still possessed immense knowledge and social influence.

Hernández immediately began hosting gatherings in his home with titicih and herbalists "who have the relevant expertise." At these colloquia, he showed them plants "so that they may help me to establish the medicinal virtues and tell me their experience with them."[27] When Mexico City officials caught wind of his activities, Viceroy Martín Enríquez de Almanza (1510–1583, who served 1568–1580) promptly ordered him to desist. Hernández responded with a plea to the Crown, asking King Philip II to command the viceroy to allow him to resume his activities. Hernández moreover successfully petitioned the king for funds to pay these native informants for their consulting services and to hire native draftsmen (tlacuiloque or *tlaquilos*) to draw the specimens. His success with Nahua doctors and draftsmen inspired him to later petition the Crown for additional funds to hire even more native "experts in your [the king's] service."[28] By inviting these indigenous physicians and herbalists into his home to discuss natural history as equals, Hernández violated local colonial protocol that insisted on the subjugation and subordination of the natives. Clearly, Hernández did not share local authorities' discomfort regarding natives gathering in groups, about interacting directly with them, or about validating their expertise.

Hernández's guides taught him first about the plants in the immediate vicinity of Mexico City, including those in hospital gardens and the old Aztec botanical gardens in Atzcalpotzalco (Azcapotzalco), Texcoco, and Huaxtepec (Oaxtepec) (see map 1), as well as window boxes and patios throughout the city. The Nahua chronicler and grandson of the Aztec emperor Moctezuma (c. 1466–1520), Fernando Alvarado Tezozómoc (c. 1525–c. 1610), described the process by which some of these plants had originally been brought to the garden in Oaxtepec, approximately 100 km south of Tenochtitlán: at the command of Moctezuma, "messengers dispersed" across the empire to collect trees and plants, "all of which were brought with their roots to be transplanted in Huaxtepec. As soon as the agent arrived on the coast of Cuetlaxtlán and established his embassy among the people of the coast, he complied by sending the trees" and plants "with their roots wrapped in *petates* [woven reed mats]." Emperor "Moctezuma delighted to see plants of such odor and color as had never been seen by the Mexica. There also arrived to plant and care for these [plants and trees] a large number of Indians [as emissaries and gardeners], numbering more than forty in addition to their wives and children." Despite their exhaustion from the journey, they immediately were presented to Moctezuma. "In front of him they" put the trees and plants into the ground, then "they made sacrifices, burned incense, and drew blood from the tops of their ears. With tears and reverence they splattered and anointed" the new plantings.[29] In 1565, Francisco Cervantes de Salazar noted that Moctezuma had "his doctors conduct experiments with" the herbs and medicinal plants in his gardens.[30] By the time of the Spanish conquest, the rendering of tribute had filled Oaxtepec and other Aztec gardens with plants from across Mesoamerica, and native doctors could draw on long experience with these, as well as with local plants, to share with the Spanish newcomer who took a lively interest in their insights.

In a letter to the king in 1521, Hernán Cortés had described the garden in Oaxtepec as "the greatest and most beautiful and fresh" he had "ever seen," and noted that it had "an infinite amount of trees of a wide variety of fruits, and many herbs and odiferous flowers."[31] In 1568, just two years before Hernández's arrival, Bernardino de Álvarez (1514–1584) had founded the Hospital de la Santa Cruz next to the garden in Oaxtepec to take advantage of its many medicinal plants.[32] Hernández was able to build on these previous efforts in fulfillment of his own mission.

Hernández's guides next led him and his retinue through fields, woods, and deserts, where villagers taught him about local simples (medicinally useful plants). Some of the people he encountered were eager to share their knowledge, while others were so suspicious of his intentions that they "could not be persuaded to reveal a single secret of nature."[33] Hernández was careful to point out to the king that he never took the natives' word for anything; he tested all their claims, along with his own hypotheses, on patients in the hospital in Oaxtepec, and even on himself.[34] From

Oaxtepec, Hernández—accompanied by his son Juan and a team of interpreters, assistants, and draftsmen—followed his guides south into the Isthmus of Tehuantepec, northwest through Michoacán and Colima, and then looped east to Querétaro north of Mexico City. Along the way, he stopped over at monasteries and the homes of local encomenderos.[35] He trusted the native guides he had recruited in Mexico City and Oaxtepec, but not the ones he hired in the hinterlands, whose "terrible lies . . . caught me off guard more than once," and even put his life at risk.[36] Throughout these travels, Hernández collected a wealth of information, and his Nahua draftsmen produced thousands of illustrations.[37]

In addition to his travels, studies of plants, and colloquia with his Nahua colleagues, Hernández gained "first-hand experience in two hospitals," learning about the use of simples and compounds and testing them on himself and on patients.[38] In 1541, the Spanish Crown had issued an idealistic decree that a hospital be founded in every Spanish and Indian town—an order that likely was ignored outside major cities.[39] Some of these hospitals were architectural masterpieces, but most were simple buildings, or even mere shacks. The few Spanish doctors in Mexico were in heavy demand by the colonial elite, so it fell to native doctors to staff these indigenous hospitals, which became spaces for the reproduction of native botanical knowledge. As medical historian Guenter Risse points out, Nahua titicih "found a temporary, albeit narrowly defined, niche among colonial medical staff, especially in New Spain's Franciscan institutions." One of the earliest of these colonial hospitals in which native doctors and staff served indigenous patients was the Hospital Real de San José de los Naturales, established by royal order in 1531 with the help of the Franciscans. In May 1553, a royal decree replaced it with the similarly named Hospital de los Naturales.[40] This was one of the hospitals in which Hernández practiced. The other was the Hospital de Santa Cruz in Oaxtepec.

In 1576, a major cocoliztli outbreak ravaged the indigenous population throughout central Mexico. It also carried off most of Hernández's titicih informants, who were on the frontlines trying to cure the sick. In the face of this crisis, Hernández stepped in to fill their shoes and to practice their methods because, as he explained, "there was no one else left in this land with any competence to" treat the dying native patients.[41] Eager to understand and contain the cocoliztli epidemic, he even hired the eminent Spanish surgeon and barber Alonso López de Hinojosos (1535–1597) to conduct clinical autopsies of the victims. Most of what we know of this tragic disease, which perhaps took a greater toll on the native population than even smallpox, comes directly from Hernández's notes and López de Hinojosos's dissections.[42] Today we still do not know for certain what cocoliztli was. But historians of science, increasingly aided by DNA analysis, point to a likely culprit: a form of salmonella introduced to Mexico by Europeans, which spread as a result of the Spaniards' destruction of the Aztec sanitation and water management systems.[43]

David Boruchoff has correctly pointed out that, though Hernández was at times "dismissive of the Nahua's grasp of medical knowledge," on account of their ignorance of the humoric theories that guided European medicine at the time, he approached "their actions from a clinician's perspective, grounding his judgment in rational, rather than moral or racial, arguments."[44] Hernández's openness to native science contrasted with other prominent Spanish writers of the era. The Creole (Criollo, that is, of European descent, born in the Americas) intellectual Hernando Ruiz de Alarcón (1574–1646), writing five decades after Hernández, for instance, rejected the common translation of titicih as "doctors." He, instead, dismissed them as superstitious charlatans in league with the devil who used "frauds and tricks" to dupe the ignorant. Writing at the same moment as Alarcón, Jacinto de la Serna (c. 1597–1681) issued a similar condemnation, while also letting on that Spanish priests and doctors in Mexico felt threatened by these Nahua competitors. He wrote: "There is no likeness so exact, nor a portrait so lifelike of a doctor that they call titzitl [*sic*], or dogmatic charlatan of the Indians, than that of a wolf, since a wolf is the color of earth, with frightening hair and varied movements. He is dirty, disgusting, lowly and cruel in his self-defense. And when out of the reach of harm, he is insolent, much as he is fearful and standoffish when he sees danger. He is a traitor, false, bloodthirsty and fond of flesh, because he is always enraged with hunger and no one can be sure of his wiles nor escape his betrayal."[45] De la Serna's tirade continues for several pages, condemning titicih less for their medical or religious ideas than for being real threats and even rivals.

Some of this difference of opinion might be shaped in part by the fact that, as Hernández informs us, most titicih died during the great cocoliztli epidemic of 1576. By the time of Alarcón and de la Serna in the 1620s, Nahua titicih might have been less rigorously trained than the ones

Hernández had met a half century earlier. Moreover, Alarcón and de la Serna, unlike Hernández, lumped into one category all titicih, herbalists, midwives, augurs, *curanderos* (medicine men), and folk religion figures. Most of their disdain, however, is not rooted in a rational critique of the titicih's methods so much as a prejudice that the colonial elites held against the colonized population.

CONCLUSION

Local colonial officials from the start were wary of Hernández, whom they viewed as a haughty interloper brandishing royal credentials. His fraternization with the natives as equals deepened local administrators' enmity toward him. These officials tried in numerous ways to place obstacles in the way of Hernández's work, forcing the protomédico to repeatedly seek the king's intervention.[46] And though Hernández acted with royal authority, even the Crown eventually became concerned about his eager collaboration with the natives. Despite the doubts and obstacles continually thrown into his path, Hernández persevered. And he had big plans. By linking his native informants' insights and methods with his own training, Hernández, as will be addressed in the next chapter, hoped to revolutionize global botany.

CHAPTER 2

TRANSLATING THE NATURAL WORLD

Unlike most of his contemporaries who wrote only with a European audience in mind, Francisco Hernández considered it imperative that he return indigenous knowledge to the natives of Mexico. In a letter to the king written in 1575, he indicated that he had hired "an Indian" to translate his multivolume natural history of Mexico, the product of his royal scientific expedition to New Spain, as quickly as possible "into Náhuatl for the benefit of the native population."

He was not the first Spaniard to translate text into Náhuatl, but his intentions were unlike anything that had come before. In 1539, Mexico's first printer Juan Pablos (c. 1500–c. 1561) had published Náhuatl translations of the bible and other religious texts to aid in conversion of the natives to Christianity. Hernández, by contrast, was not trying to convert the Nahua, but rather to safeguard their knowledge and return it to the local people. He also seems to have been trying to solidify his legacy among a group of people whose opinions most Europeans considered unworthy of consideration. The *protomédico* placed such importance on this translation that he declared to King Philip I that until it is done the expedition to Mexico "remains incomplete."[1]

Hernández was well aware that the smallpox outbreak of 1519–1520 and the *cocoliztli* outbreak of 1545 had decimated the population and fragmented indigenous medical and botanical knowledge. His commitment to completing the translation of his findings, gathered thanks to his indigenous colleagues, into the native language soon became all the more pressing when he saw the cocoliztli outbreak of 1576 claim the lives of 50 percent of the remaining indigenous population in central Mexico.[2] Despite his efforts to study the disease and treat the sick, Hernández felt helpless in the face of cocoliztli's devastation, watching in horror as the pathogen wiped out the native *titicih* and *tlacuiloque* whose expertise filled the pages of his books. Sadly, Hernández's Náhuatl translation has not survived. Yet, the fact remains that, at a time when few scientists bothered translating their Latin texts even into any European vernacular, Hernández's insistence on making it accessible in the language of a conquered people was extraordinary.

By reconstructing the taxonomic and visual systems that the protomédico and his native collaborators created, this chapter argues that Hernández did not just want to preserve native knowledge, but to improvise on it to create the basis for a new system for naming and depicting nature, and then to insert this new system into the heart of European science as well as into local practice. Historian Stuart McManus has argued that "Renaissance humanism was not one thing. Rather, it was a versatile toolbox of ideas, linguistic techniques, and scholarly practices that could be put to multiple ends by humanists trained in its methods."[3] The improvisations on the part of both Hernández and his Nahua collaborators expanded this Renaissance toolbox while cementing novel understandings of the relationship between plants, place, and culture within Mexico.

SIFTING THROUGH VERSIONS OF VERSIONS OF VERSIONS

Upon his return to Spain in 1577, Hernández expected to be celebrated as a hero for his contributions to natural history. Instead, members of the royal court expressed concerns about how he had carried out his task, then removed him from his job as protomédico. The king replaced him with Nardo Antonio Recchi (1540–1594), who in 1580 began going through Hernández's manuscripts to compile a selection of the plants that he judged to have the greatest medicinal or economic potential. Recchi ended up including only 12 percent of the plants Hernández had studied.[4]

Finding Hernández's organization strange and incomprehensible, he completely reorganized the items that he selected according to typical European conventions of the time. Recchi also removed all Hernández's images, which seem to have struck him as even stranger and more incomprehensible than the text, and replaced them with new drawings that followed European conventions. Hernández's original texts and images were seen by only a handful of people within Europe before they were squirreled away by crown officials, and then incinerated during a fire that tore through the royal palace in 1671. Soon almost all that remained of Hernández's finding were the distorted glimpses provided by Recchi.

With good reason, historian Raquel Álvarez Peláez has criticized Recchi for incompetence in his treatment of Hernández's text. Germán Somolinos goes further, accusing Recchi not only of butchering the organization of Hernández's texts, and hence destroying his intentions, but also of doing a mediocre job of summarizing the parts of the text that he kept.[5] While some scholars have tried to rescue Recchi's reputation, most have tended to agree with Álvarez Peláez and Somolinos. Even the most charitable of accounts recognize that Recchi profoundly distorted what Hernández was trying to accomplish, and, as a result, made his findings harder, rather than easier, to digest and integrate into European Renaissance science.[6]

Before it appeared in Europe, Recchi's manuscript was published without images in Mexico in 1615, translated into Spanish and edited by Francisco Ximénez (c. 1560–1620). Ximénez's version, called *Quatro libros de la naturaleza y virtudes de las plantas y animales que estan recividos en el uso de Medicina en la Nueva España* (*Four Books on the Nature and Virtues of Plants and Animals Obtained for Medical Use in New Spain*), was published by María de Espinosa López Dávalos (c. 1560–c. 1616), who ran the print shop founded by her late husband Diego López Dávalos (b. unknown–c. 1612).[7] This version later gained influence within Mexico during the late nineteenth and early twentieth century as a means of accessing the findings of Hernández. When Gabriel Alcocer wrote his botanical study of mesquite in 1903, for example, it was Ximénez's work that he consulted. According to Alcocer, Hernández "was the first to write about the uses the Indians made of mesquite, but since [a version of] his work was not published until 1694 in Rome, then in 1790 in Madrid, one could say, more exactly, that Father Francisco Ximénez, the translator and annotator of Hernández, who published his work in Mexico in 1615, was the first to publicize information about mesquite."[8]

Some European and Mexican naturalists became familiar with Hernández's work either through Recchi or else Ximénez's version of Recchi's work, but most learned about it through a version of Recchi's work created by Federico Cesi (1585–1620). In 1603, Cesi created the Accademia dei Lincei in Rome, often considered the first modern scientific academy and which was responsible for promoting the use of the microscope in the study of botany.[9] Cesi purchased a manuscript copy of Recchi's version of Hernández's findings. Inspired by what he saw, he placed the deciphering, analysis, revision, and eventual publication of a version of Hernández's findings at the center of the Accademia dei Lincei's efforts. In 1628, the Accademia published the massive 950-page *Rerum medicarum Novae Hispaniae thesaurus*, which was the Linceian (that is, produced by the Lincei Academy) version of Recchi's version of Hernández's version of what the titicih had told him. This 1628 version of a version of a version became the medium through which most European scientists studied New World botany.

Through the various revisions at the hands of men such as Recchi and Cesi, and its passage through versions of versions of versions, Hernández's research was stripped of the organization he had used, all the images his tlacuiloque had produced, and all but a fraction of the plants he had studied. What European naturalists could not easily strip from his manuscripts were the Náhuatl names that Hernández had assigned to the plants. But instead of fully accepting these Náhuatl names, or the nomenclature system they were part of, they simply added them to the growing profusion of Latinate and vernacular names that the New World plants had begun to acquire.

Taxonomy and images were among the elements that Hernández most eagerly adapted from his Nahua collaborators and can offer some insight into the process of cultural exchange regarding the study of nature. The rest of this chapter, therefore, focuses on recovering and analyzing Hernández's system of nomenclature and the botanical images that were stripped from his manuscripts. It considers what these names and images can reveal to us about emerging understandings of indigeneity and nature in Mexico, and suggests why and how his efforts inspired and shaped initiatives that would unfold in the coming centuries.

A "METHOD FOR IDENTIFYING THE PLANTS OF BOTH WORLDS"

Hernández's medical training at the University of Alcalá made him well aware of the problems caused by the profusion of plant names of his era, which were later mocked by Rousseau (see chapter 1). This made him receptive to the benefits of the Náhuatl nomenclature system he learned in Mexico, which he adapted as the basis for his own taxonomic method. In letters to his friend Arias Montano and to King Philip II, Hernández bragged that he had written a manual titled "Method for Identifying the Plants of Both Worlds," in which he described a system that he had devised "for identifying Indian plants as well as ours." He applied this method in another book he had written, which he summarized as a compendium of "the plants native to this world and the virtues attributed to them by the natives."[10] No fragment of either book survives, nor do we have any detailed secondary account of their contents. Yet we do have hints into how his method functioned. These hints can be found in the nomenclature that Hernández applied in his research and in comments by him and others regarding the manner in which he had organized his findings.

Despite his respect for Nahua botanical knowledge, the protomédico criticized his titicih collaborators for not following European humoric theory, particularly their lack of interest in surgical interventions such as bleeding to balance the humors. "They never cut anyone's veins," he complained. "Wounds are treated with simple medicines or by covering them with flours; with these things they are usually helped, but" without humoric theory to guide them, the titicih applied cold when Galen says they should apply heat, and heat when they should apply cold. The titicih responded "haughtily" when Hernández tried to correct them by introducing them to the teachings of Galen.[11] His concerns about their ignorance of humoric theory did not stop him from praising their knowledge of plants and their plant nomenclature system. "It seems amazing," he wrote, "that among such uncultivated barbarous people" all the plant names are "adapted . . . with such precision and care, that the name alone is enough to indicate the nature of any important thing."[12] Jakob Dietrich and other Renaissance botanists dealt with the confusion that surrounded their specimens by listing as many Latinate and vernacular names as possible for each plant they studied (see chapter 1). But Hernández felt that a universal system adapted from Náhuatl might be a more efficient solution.

Hernández did not simply apply existing Náhuatl vernacular names to the plants he studied, as scholars sometimes incorrectly presume. As historian of science Paula de Vos has shown, indigenous Mexico experienced as much chaos and inconsistency in the naming of plants as did Europe, and so simply adopting Náhuatl names would not have solved science's nomenclature woes.[13] During his travels and through his conversations with native informants, Hernández would have been very aware that indigenous vernacular names, in their raw form, would not eliminate the chaos of nomenclature that confounded naturalists. So, instead of simply borrowing an existing method of taxonomy, Hernández took on the task of using Náhuatl's linguistic structure as the basis for the creation of *new system* of nomenclature and taxonomy (that is, a system both for naming and organizing plants) that was more stable than existing European systems in terms of having a name and category reliably attached to a particular kind of plant. Because this new system was based mostly on Náhuatl, it would be capable of assimilating New World plants.

Rather than simply adopt Náhuatl plant names, Hernández sought to build on some of the native system's virtues to devise a stable taxonomy that he hoped could resolve nomenclature in both Europe and Mexico. Simon Varey has shown that Hernández's botanical descriptions are "obviously European in conception," based on the theories of Galen and other ancients rather than on "Aztec conceptions of illness, life, and death," but that the protomédico had sought to "arrange his material" according to a taxonomic structure that "was entirely Mexican, not European," and which would replace the existing Theophrastian categories of trees, shrubs, and herbs.[14] Hernández argued that this system should not just be used for New World plants, but also adapted for those from the Old World, and even for the plants that Europeans were encountering in other parts of the globe, from Peru to China. This was a radical proposition, but we can see how it might have seemed quite reasonable to Hernández, who was about to introduce Europeans to three thousand new species, increasing the number of known plants six-fold.

Recchi found Hernández's Náhuatl-based taxonomic structure unsettling and confusing, so he intentionally purged it from his interpretation of the protomédico's work. Science historians José López Piñero and José Pardo Tomás have pointed out that Recchi tried to rearrange his selections from Hernández following the traditional Theophrastian categories that

Hernández had rejected: "Thus, Recchi eliminated at a stroke Hernández's taxonomic innovations" that were "based on Amerindian terminology."[15] Ripped out of the system of organization that Hernández had devised, the names and implied connection among his plants became confusing, almost undecipherable. Though it was Recchi's revisions that caused the resulting confusion, he and other Europeans blamed Hernández for a supposed lack of organization.

So, what was this Náhuatl-based nomenclature system that Hernández tried to promote? Because Hernández's "Method for Identifying the Plants of Both Worlds" has been lost to us, and Recchi tried to purge Hernández's botanical manuscripts of the titicih's influence, we can't know the details with certainty, but we can get a general idea. Medical anthropologist Bernardo Ortiz de Montellano and historian of science Raquel Álvarez Peláez have reconstructed aspects of how it might have operated.[16] Ortiz de Montellano offers willow as one of his examples. The general term for willow in Náhuatl was *huexotl*, which would be equivalent to what we now might consider the genus. This was divided into several categories that we now might consider somewhat on the level of species, such as *ahuexotl* (aquatic willow) and *quetzalhuexotl* (regal willow), denoted by adding a prefix. Each species was then further divided into what we today might consider varieties. One variety of ahuexotl, for instance, was *ahuexocaquilitl* (aquatic edible willow), denoted by the addition of a suffix. In this example, we can see how the main root huexotl is carried into each category. By adding "*atl*" (meaning water), shortened to "*a*" to the start of the word, we get the species ahuexotl. The variety receives the suffix *caquilitl* (meaning edible) to form a[tl]+huexotl+caquilitl = ahuexocaquilitl. The various roots, prefixes, suffixes, and modifiers evolve in a predictable manner, conveying information about where and under what conditions a plant grows, what it looks like, and what properties or uses it has. This made it more efficient than European vernacular or Latinate names at identifying plants, because it decreased the chances of accidentally applying the name to some other plant, unless that other plant happened to share the same combination of characteristics.

Rather than moving radically from one vernacular name to another (as did most European folk names), or cobbling words together with arbitrary signifiers (as we will see later during the Enlightenment with the emergence of Linnaean taxonomy, when many plants were assigned Latinized names of prominent male European scientists), Náhuatl constructed the plant and animal names by combining roots and prefixes in a fairly predictable way, with names that had the additional virtue of conveying useful information about the plant. It is easy to see why this appealed to Hernández.

As soon as Hernández began to grasp the Náhuatl structure for naming things, he introduced his own modifications, assigning his own suffixes and prefixes. In the end, he seems to have assigned names that were not necessarily those used by the Nahua people, or else he chose as the official designation a name used in one place and ignored the name used in another location. It is difficult to decipher from his text which suffixes and prefixes came from Nahua informants and which were added by Hernández. Nor is there a way to determine when or why he might have preferred the name from one region over that from another.[17]

This has frustrated Ortiz de Montellano and others who try to reconstruct aspects of Nahua culture before the conquest, but for my purposes it reveals the extent to which Hernández was not simply reproducing an existing system, but altering it to fit European needs and establishing his voice of authority as the supervisor of this system (much like the role Linnaeus would assign to himself two centuries later). Hernández thus improvised on the Náhuatl structure to create something novel, yet recognizable, and capable of assimilating all the world's plants. We can also imagine why European scientists who had no knowledge of Náhuatl, and who perhaps looked down on native cultures of the Americas, rejected it.

ILLUSTRATING NATURE: THE VISUAL LEXICON OF NATIVE DRAFTSMEN

The team of experts that Hernández assembled on his arrival in Mexico City comprised not just native doctors, but also native draftsmen known as tlacuiloque. In the first few months alone, his team of indigenous artists illustrated over eight hundred "new plants, never before" known to European naturalists, and would go on to paint over three thousand botanical illustrations.[18] Hernández esteemed these draftsmen and their technique of representation so highly that, when the king asked him to extend his expedition into Guatemala and Peru, he insisted that he would do so only if he could bring along his Nahua tlacuiloque.[19] He even included his native collaborators in his will. After giving his soul to God, but before mentioning any of his children, Hernández allotted an inheritance to the titicih, tlacuiloque, herb collectors, and the numerous other Indians "who worked in this enterprise," with special mention of and generosity to his three favorite tlacuiloque, whose Christian names were Pedro Vásquez, Baltasar Elías, and Antón (Elías?).[20]

During the expedition, Hernández informed his king that the native draftsmen he had hired "to paint the natural things" were meeting his scientific needs "quite well."[21] Given the importance Hernández placed on images and his esteem for the draftsmen's work, we must ask what kinds of images they produced and what he valued about their strategies. We also need to consider what these tlacuiloque's images, and the process of negotiation between Hernández and the artists, might reveal about perceptions of Mexican nature on the part of Hernández and on the part of the Nahua. To answer these questions, we must first address what kind of training these draftsmen might have had and what kinds of images of plants native tlacuiloque were producing around the time of Hernández's arrival.

Elizabeth Hill Boone reminds us that, until the conquest, "painted books pervaded almost all aspects of Aztec life," but during their political and spiritual conquest of Mexico, the Spanish incinerated almost all of them.[22] Codices from between the time of the conquest and Hernández's 1571 arrival are almost equally scarce. Fortunately, one remarkable text has survived: an herbal from 1552 attributed to two native scholars, the ticitl Martín de la Cruz and the Latin linguist Johannes Badianus.

The book's official title is *Libellus de medicinalibus Indorum herbis* (*Booklet of Indian Medicinal Plants*), but it is more commonly known as the *Codex Cruz-Badiano*. It was presented by Francisco de Mendoza (1523–1563), son of Viceroy Antonio de Mendoza (c. 1490–1552, served 1551–1552), as a gift to the king of Spain, the Holy Roman Emperor Charles V. Its purpose seems to have been to convince the Crown of the Indians' intelligence, to secure funding for the Colegio de Santa Cruz de Tlatelolco. After delivery to the Crown, the herbal was deposited in the royal library and not mentioned again for almost one hundred years. It then was studied by the Accademia dei Lincei and eventually made its way into the Vatican Library, where it remained until Pope John Paul II repatriated it to Mexico around 1990.[23] The book measures only 6 inches by 8 inches, but is richly illustrated with between 184 and 204 full-color botanical paintings (the number depends on what we count as an individual illustration), and is the earliest known herbal from the Americas.

We know little about Martín de la Cruz or Johannes Badianus, except the following: (1) they were indigenous men who were likely already young adults at the time of the conquest; (2) Badianus originally was from Xochimilco, just south of Mexico City; (3) Martín de la Cruz's original name might have been Momahuti; and (4) they converted to Christianity, studied under the Franciscans, and eventually became instructors at the Real Colegio de Santa Cruz de Tlatelolco, where De la Cruz taught native botany while Badianus taught Latin.[24]

No records survive, beyond the words and images within the book itself, regarding how the authors went about their work of collecting plants, their methods of analysis, or their goals. But I propose that careful analysis of the images, and correlation with the historical context, can reveal a surprising amount about how Nahua intellectuals in that era thought about botanical illustration, and why Hernández would have been enthusiastic about their approach.

The Real Colegio de Santa Cruz, where De la Cruz and Badianus taught, was built on the same site as, and literally from the rubble of, the calmecac that had educated the children of the Mexica elite before the conquest. In 1526, five years after the defeat of the Aztecs, Pope Clement VII ordered that twenty-six children of the Mexica nobility travel to Spain to be educated and then return to Mexico to instruct their people. Rather than carry this out, the Franciscans in 1533 gathered fifty children from among the indigenous nobility of the central valley to provide them with a local multilingual instruction—Náhuatl, Spanish, Latin, and Greek—as well as Christian theology, alphabetic script writing, singing, music, sculpture, and painting/illumination. Three years later, in 1536, the viceroy endorsed the Franciscans' effort by authorizing the founding of the Colegio de Santa Cruz, and was soon seconded by the king of Spain.[25] With royal support, the number of students tripled and the curriculum expanded to include philosophy, logic, arithmetic, and native botany and medicine. When members of the first generation of Christianized indigenous students, including De la Cruz and Badianus, graduated, they took over most of the teaching and administrative responsibilities of the school.[26]

The Crown later reversed its liberal policy toward the preservation of indigenous culture. In 1600, it forbade Indians from occupying educational posts and required that institutions teach exclusively European and Christian customs.[27] In 1605, the school lost its government support and in the mid-seventeenth century was abandoned. Its decline also was precipitated by the multiple waves of cocoliztli and other plagues that killed most of its teachers and students. Despite its sad fate, the Colegio de Santa Cruz was a thriving center of learning at the time De la Cruz and Badianus completed their herbal and when Hernández visited Mexico City in the 1570s. Franciscan missionary Bernardino de Sahagún, one of the teachers at the school, even relied on its graduates in the preparation of his famous *Florentine Codex* (completed around 1576).[28]

Scholars generally presume that Martín de la Cruz drew the images in the herbal, but this is probably incorrect. Emily Walcott Emmart's study of the inks, the strokes of the lines, and other details of the herbal revealed that they were produced by a single hand. She presumed that hand belonged to De la Cruz, and later scholars have accepted her assumption.[29] Titicih were trained in botany and in the practice of medicine, while tlacuiloque were trained to read and produce glyphic codices and, after the conquest, alphabetic manuscripts. Just as a Renaissance artist in Europe would produce his own pigments, sizing, and brushes, so a *tlacuilo* had to master the crafts of making paper, producing pigments, manufacturing brushes, and, increasingly, bookbinding. There is no basis for thinking that De la Cruz, a doctor and teacher, and probably also an attending physician at the Hospital de los Naturales, would have had the training or time to produce the paints and execute the illustrations. It is equally unlikely that Badianus, an expert in Latin, not botany or art, would have drawn the illustrations.

The argument that the authors and the artist were different people is strengthened by the fact that the text and the images don't relate to one another.[30] The Latin text, which is aligned with Pliny and other European sources of authority, explains the humoric causes of illnesses and specifies how to prepare and apply each plant as treatment for physical and psychiatric ailments. It makes no reference to the visual images, which appear to rely on native visual practices and make no attempt to depict what is described in the text. All this suggests that the images almost certainly were rendered by a different person than the narrators of the text. The hand we see is likely that of a specially trained tlacuilo.

What does the *Codex Cruz-Badiano* teach us about how a tlacuilo of the era would have gone about his work of depicting nature? Studies of the *Codex Cruz-Badiano* tend to either dismiss the book as "contaminated" by European ideas, because its text draws on humorism, or else they romanticize it on account of its use of Náhuatl names and native visual traditions.[31] Medical anthropologist Bernardo Ortiz de Montellano, for example, dismisses the book because of its "European contamination," as evidenced by its text. His concern is understandable, if we recall, as mentioned earlier, that his goal is to understand preconquest Aztec medicine, something that is hard to do with this text. Even the art historian Donald Robertson, in his classic study *Mexican Manuscript Painting*, is quick to dismiss the codex on the basis of his claim that the book's text renders it insufficiently authentically native to be of interest.[32]

Art historians Carolyn Dean and Dana Leibsohn have correctly argued that scholars too often presume that they "have to see pre-Hispanic forms to be convinced that" an object can "be recognized as indigenous. This denies the radical transformations in the lives of indigenous peoples brought about as a result of colonization. Such interests also betray desires to freeze indigenous people" in a particular past moment.[33] Building on the insights of Dean and Leibsohn, I propose that, rather than judge the codex on the basis of its supposed level of cultural purity, or lack thereof, we instead consider what its images can teach us about how Nahua draftsmen navigated among shifting strategies for visualizing the products of nature.

Taking this approach, I also argue that the codex can offer insight into the methods embraced by Hernández and his native tlacuiloque collaborators two decades later in the 1570s. In particular, it can reveal what native draftsmen saw as important about a plant, how they understood the plant's relationship to its ecological and cultural context, and what strategies they used to encode such information within the image. Critical examination of the codex, therefore, can help us understand what intrigued Hernández about native draftsmen's methods, and how he made use of their approach.

Most scholars who study the *Codex Cruz-Badiano* focus their attention on the text. Those who do pay attention to the images often dismiss them as quaint, vague, and unduly abstract.[34] However, if we open ourselves to the possibility that the *Cruz-Badiano* draftsman was operating under a different idea than our own regarding how visual images convey information, then we can ask what the tlacuilo sought to convey by employing particular strategies.

The meaning of visual accuracy is less straightforward than one might presume. Illustrations we have come to think of as scientifically accurate were, in fact, never meant to directly depict what the artist saw with his own eyes. Enlightenment-era botanical illustrations seem extremely accurate to our modern eyes. Yet, if we were to compare one of these illustrations to a sample of the plant species it represents, we would find that they cannot square up. Starting in the Enlightenment, draftsmen were meticulously trained to produce a false sense of visual mimeses; they based each drawing on what botanists insisted a species of plant should look like in an ideal world, not on what the artist directly observed from a particular specimen. Draftsmen manipulated their images in dramatic ways, exaggerating some features, downplaying others, and sometimes removing elements altogether, or adding visual elements that are not even present in the particular specimen they were using as their model.[35] Our belief in the accuracy of such images is based on our acceptance of certain conventions that became normalized over time, and by our presumption that precision of line indicates verisimilitude. We must ask, then, what alternative strategies might have guided the eyes and hands of Nahua tlacuiloque in sixteenth-century Mexico.

A striking aspect of the images in the *Codex Cruz-Badiano* is the difference between how the tlacuilo has depicted the upper portion of the plants as compared to the roots and the substrate (see figure 2). The top parts of the plants are drawn in a way that would have looked relatively familiar to European illustrators of the Renaissance era. The lower portions of most of the plants, however, depart markedly from what one would have seen in Europe. Given how intentionally the Latin text and the overall organization of the book speak to recognizably European conventions, we can surmise that the tlacuilo could have chosen to depict the lower part of the plant in a more European manner had he wished to do so. The departure from recognizably European conventions, therefore, likely resulted from deliberate decisions.

The question then becomes why the draftsman would have decided to depict the lower part of many of the plants in this seemingly unusual manner. What did this strategy of depiction offer that a conventionally European mode did not? In the absence of textual explanations by native

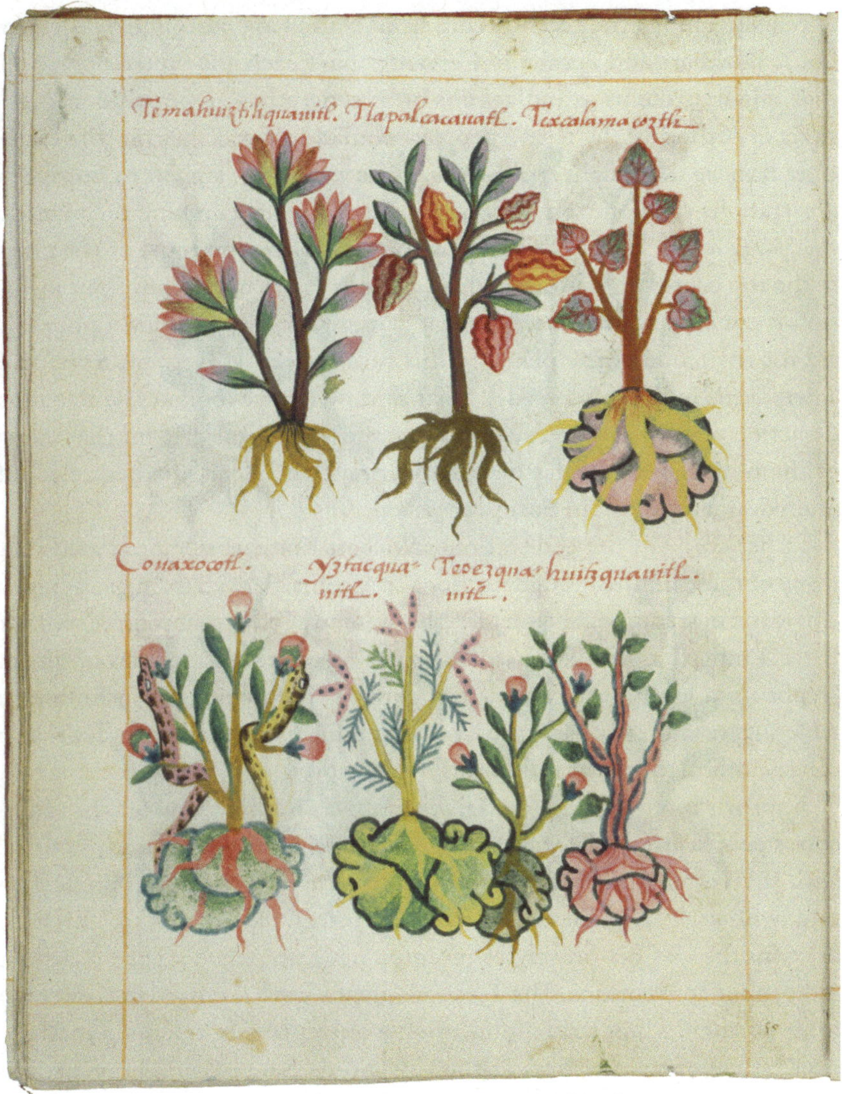

FIGURE 2 *Codex Cruz-Badiano*, page 38 rev (1552). Courtesy of the Instituto Nacional de Antropología e Historia, Mexico City. https://codices.inah.gob.mx/pc/.

titicih and tlacuiloque regarding the illustrations, we must use whatever other clues are available to try to decipher the images.[36]

When looking at these images, we must be careful not to take every unfamiliar visual device as an act of cultural resistance. In their studies of religion among the recently converted Nahua in postconquest Mexico, Louis Burkhart and Fernando Cervantes have each shown that practices that might strike us as indigenous or as syncretic were, instead, expressions of orthodoxy by native people.[37] Similarly, details within the *Codex Cruz-Badiano* that might strike us as exotic were more similar to European illustrations than one might presume. The serpents climbing the plant to the lower left in figure 2 or the glass box enclosing the roots of the plant in the top center of figure 3 might strike our eye as exotic, but their inclusion is consistent with the European convention of providing a phonetic aid for the name of the plant (in this case, *couaxacotl* [serpent fruit] and *acamallotetl* [imprisoned reed]). That said, when we look at the tlacuilo's depiction of the roots and substrates of the plants, it is evident that many of them display strategies of visualization that depart deliberately and often dramatically from European practice.

In her study of the codex, Emily Walcott Emmart demonstrated that many of these unusual root depictions are glyphs, derived from pictographs employed in traditional Nahua writing systems. To try to interpret some of these, Emmart assembled a table comparing water-color copies of fifteen glyphs used within the codex (see figure 3). The depictions in Emmart's table suggest the creative approaches that indigenous draftsmen devised to convey information that Europeans might express only in alphabetic script.

Improvisation within pictorial systems was nothing new for the indigenous people around Mexico City at that time. They were already familiar with the Testerian alphabet introduced by the Franciscan Jacobo de Testera, who arrived in Mexico in 1529. To introduce Indians to Christian doctrine, he and his proselytes invented new, simplified glyphs based on indigenous conventions. The Testerian glyphs were arranged in sequential form so that, when read aloud, they would provide religious instruction.[38] One of the teachers at the Colegio de Santa Cruz, Fray Toribio de Benavente (1482–1565), known as Motolinia, described a scene that suggests indigenous people quickly became comfortable with this kind of pictorial improvisation:

> One Lenten season . . . the number of those who came to confess was so great that I was unable to give them the advice I should have liked to give them. I told them that I could hear confession of only those who would bring their sins written down in figures, because writing in figures is a thing

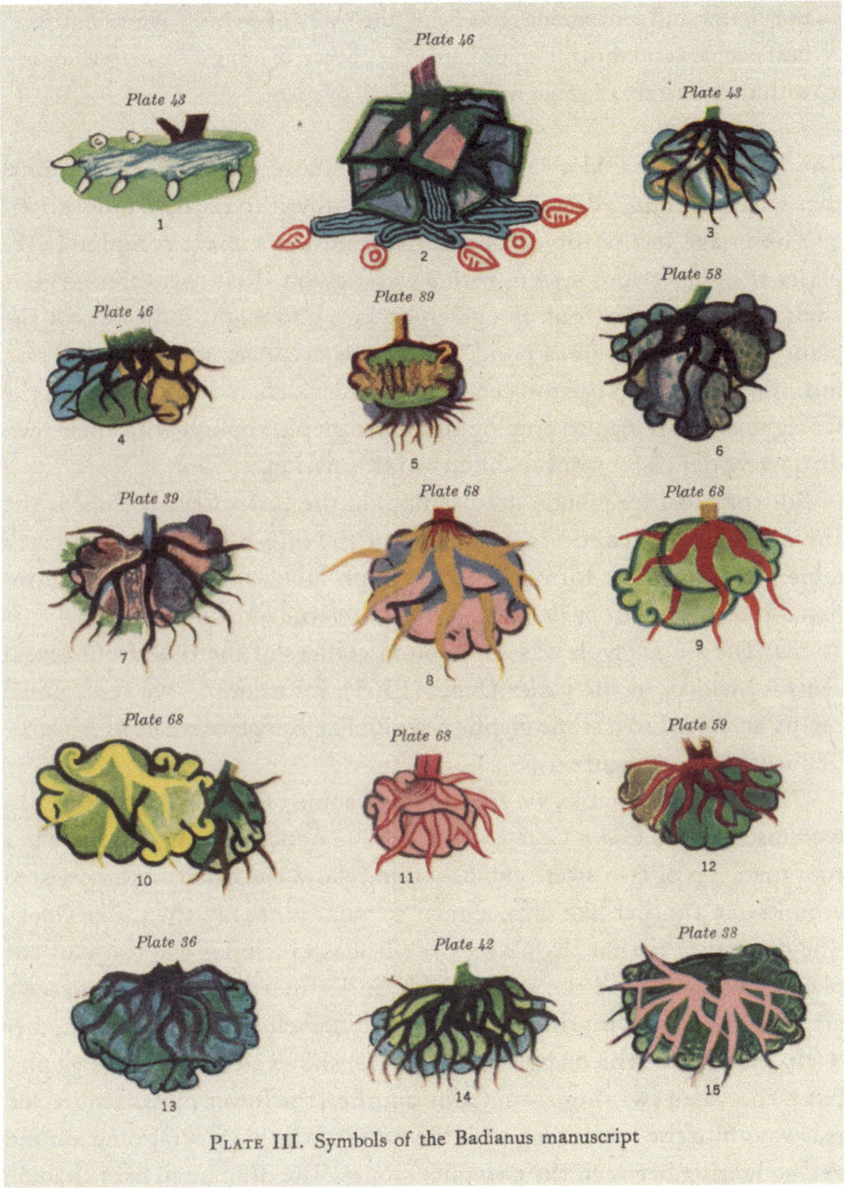

FIGURE 3 Table comparing roots in *Codex Cruz-Badiano*, from Emily Walcott Emmart, "Introduction," *The Badianus Manuscript: Codex Barberini (An Aztec Herbal of 1552)* (Johns Hopkins Press, 1940), 35, Amherst College Archives and Special Collections.

> they know and understand, this being their way of writing. It was not to deaf people I said this, for immediately they began to bring so many writings with their sins that I could not attend to all of them.[39]

The glyphs in the *Codex Cruz-Badiano* were more formal and traditional that what Motolinia describes, but no less subject to creative innovation.

The image in the top-left corner of Emmart's chart completely replaces the plant's root system with a water glyph. That the water glyph is elongated and set in front of a green background might indicate that the plant grows in a stream, a pond, or perhaps even wet soil. Other glyphs, including the one in the top center of Emmart's chart and the one on the far right of figure 6, also employ the water glyph, though with their own distinctive details, to convey different information.

But the most frequently used symbol in the *Codex Cruz-Badiano* is the stone glyph, which appears in thirteen of the fifteen images in Emmart's table. In its simplest form, the stone glyph looks something like an inflated rectangle with curled edges, but can come in a wide range of variations. The stone glyph was common in codices of the time from across central Mexico. In the *Codex Osuna* (1565), for example, we see a nopal cactus anchored to a stone glyph to symbolize the place name of Mexico-Tenochtitlán (see figure 4).

Using Emmart's table, we can compare some of the ways stone glyphs were used in the *Codex Cruz-Badiano*. At the upper right of her table is a root made up of two stone glyphs—one yellow, the other blue—twisted around one another like taffy, similar to what we see in the *Codex Osuna*. The difference is that the *Codex Cruz-Badiano* includes the roots of the plant flowing around the stone glyph. In Emmart's number 4 we see a single stone glyph with three color zones: blue, yellow, and green. Number 8 (the original of which appears in figure 2) shows two pink stone glyphs, but rather than twisting around one another, the lower one is embedded sideways into the upper one, with the roots of the plant wrapping around and squeezing between the two pink stones. The draftsman uses shading to help differentiate the two glyphs, and to emphasize how they relate to the plant's root structure. The green cluster in number 10 (the original of which appears in figure 2) is more complicated. It incorporates a total of five different stone glyphs anchoring two different plants, one with yellow roots, the other with brown. Each stone glyph is green with dark outlines.

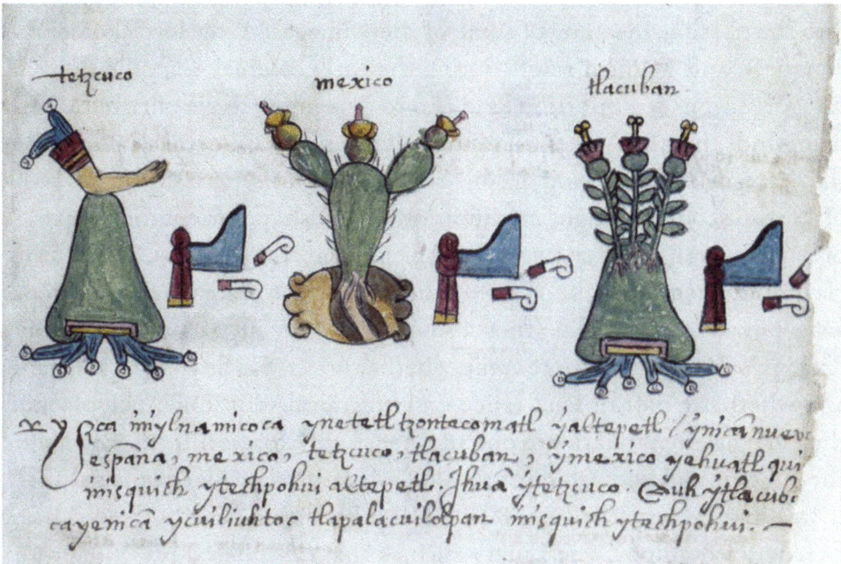

FIGURE 4 *Codex Osuna* c. 1565, folio 70, U.S. Library of Congress, World Digital Library.

The cluster anchoring the plant to the left is composed of three glyphs, while the cluster to the right is composed of two dark-green stone glyphs, one sitting atop the other, with roots growing around and between them. These clusters might indicate that the two plants grow in a linked substrate, a single ecosystem, or else that they are symbiotic. It is hard to know precisely how we should interpret all this imagery, but close observation and comparative analysis might offer clues.

Art historian Elizabeth Hill Boone argues that pictographic images in indigenous codices often encode information about place, time, events, qualities, and even abstract concepts.[40] Sometimes, the information conveyed by a plant's root glyph in the *Codex Cruz-Badiano* might seem relatively straightforward, as with the hard seed contained within the root of the *tlalcapol*, or earth cherry. Other times it can become complicated.

While we can sometimes guess at the kind of information the images are designed to convey, we lack the sources we need to interpret them with precision. An intriguing advance has come from geographer-geologist Barbara J. Williams. By comparing numerous codices and land documents, as well as contemporary and historical soil analyses, Williams offers insight into the possible meaning of some of these images. Aztec lords before the conquest, and Spanish rulers during the early colonial era, relied on soil categorizations to apportion land and to determine levels of tribute. Williams finds that the Aztecs used stone glyphs to describe soil types because they categorized each kind of soil, from dry to mesic, partially on the basis of its stone content. Slight modifications in the shape, orientation, or color of a stone glyph could show the kind of soil and its variations, as well as its geographical situation, such as whether it was on a plateau, a gentle slope, a steep incline, or a basin. They also could convey information about hydrology, such as moisture content, whether water was flowing or stagnant, and whether a plot of land relied on irrigation. Additionally, glyphs and their arrangement relative to one another could express information about the content and ecosystem of the soil where a particular plant thrives.[41] It was natural that the *Cruz-Badiano* tlacuilo would have borrowed from a recognized graphic vocabulary such as that deciphered by Williams to convey information about where and how plants grew.

In many of the images the draftsman dispensed with traditional glyphs to either invent his own symbols or to otherwise depict context. In some he draws mounds of rock that seem to indicate rocky outcroppings or cliffs. In others, such as those in figure 5, he uses a halo of color and other visual clues to indicate the kind of substrate in which the plant grows or other characteristics of the plant. In figure 6 we can see how he modifies the traditional water glyph by applying it along the length of a plant's roots.

Williams confirms that the tlacuilo who drew the images in the *Codex Cruz-Badiano* seems to have been improvising rather than strictly following traditional conventions. Given her goal of understanding what the soil was like in specific locations at particular historical moments, Williams characterizes such invention as a weakness of the document. But, here again, for my purposes such improvisations prove particularly revealing. They remind us that, rather than evaluate Hernández's Náhuatl nomenclature or tlacuiloque's botanical illustrations on the basis of their supposed levels of indigenous purity or corruption, we can instead recognize them as part of the process of cultural encounter. The botanical illustrations in the *Codex Cruz-Badiano*, therefore, should be recognized as innovative approaches to botanical representation that attempt to bridge Nahua and European conceptual practices.[42]

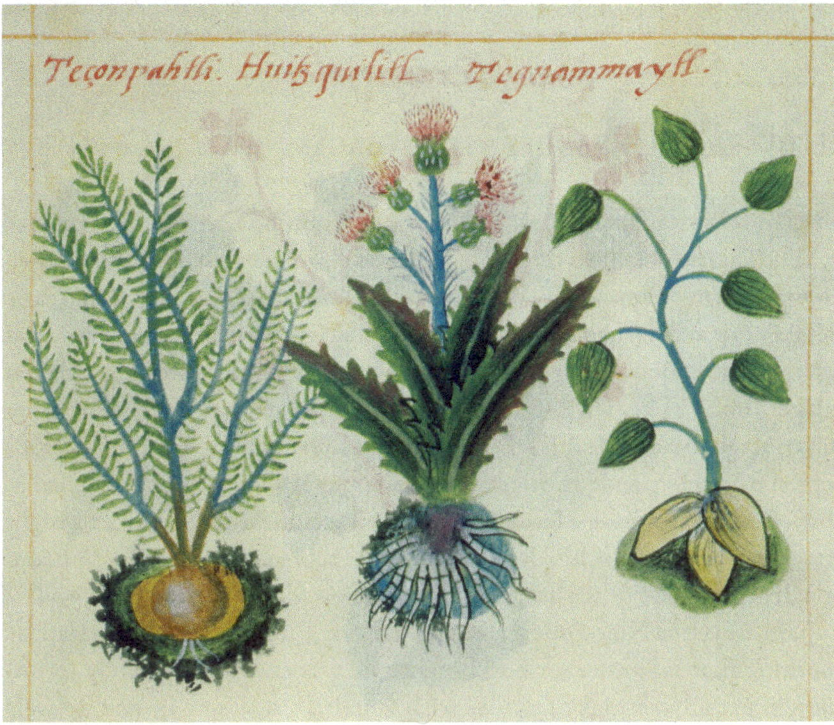

FIGURE 5 *Codex Cruz-Badiano*, page 8 rev (1552), Instituto Nacional de Antropología e Historia, Mexico City. Courtesy of https•//codices.inah.gob.mx/pc/.

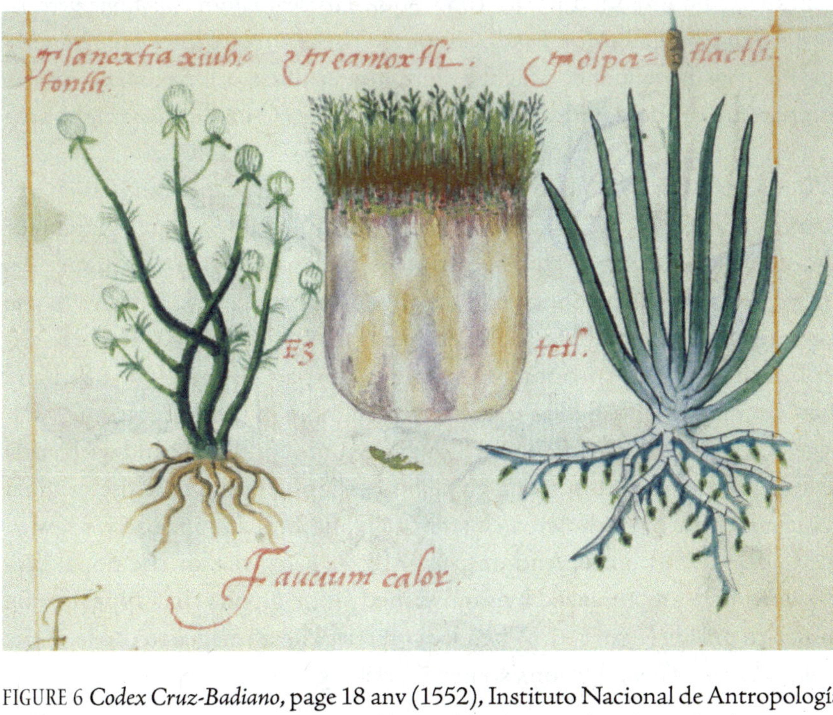

FIGURE 6 *Codex Cruz-Badiano*, page 18 anv (1552), Instituto Nacional de Antropología e Historia, Mexico City. Courtesy of https•//codices.inah.gob.mx/pc/.

RECOVERING HERNÁNDEZ'S LOST IMAGES

Hernández's draftsmen likely had the same training and deployed similar methods as the tlacuilo who painted the images in the *Codex Cruz-Badiano*—Hernández might even have employed the same tlacuilo as part of his team. While none of Hernández's illustrations survive, we do have written, and even visual, evidence that can inform us about what a few of the original images looked like. I will examine two kinds of evidence. First, we have three images that the Jesuit scholar Juan Eusebio Nieremberg (1595–1658) copied from Hernández's original manuscripts in Madrid and published in 1635. Second, we have Hernández's own descriptions. These offer insight into what he made of his draftsmen's approach to visual production and how he thought it fit within his broader vision for his expedition.

Juan Eusebio Nieremberg, who never left Europe, was fascinated by anything that was novel or odd because he was convinced that by reconciling these aberrations of nature with Christian doctrine he could reveal underlying moral principles.[43] This search for oddities is what led him to the botanical paintings in Hernández's text, which he saw first-hand in Madrid. He had three of these copied and made into black-and-white prints that he included in his 1635 book *Historia naturae maxime peregrinae* (*A History of the Most Foreign Nature*). These three prints are the only surviving visual evidence of what Hernández's illustrations looked like. A comparison of these with the images in the *Codex Cruz-Badiano* might help us gain an idea of Hernández's originals.

On page 306 of his book, Nieremberg shows *Atatapalacatl*, which appears as a water lily whose roots have been replaced by a water glyph (see figure 7), similar to what we saw in the *Codex Cruz-Badiano* (figure 3). He explains that Hernández had described this water lily as native to the lakes around Mexico City, where it grows in slow-moving water. He describes the roots as sprouting from the lower part of the plant-like hairs that "become established in the mud, and almost in the water itself."[44] On page 308, we see *Teoamatl* (figure 8), whose roots cover a mound with curls that recall those used in stone glyphs—perhaps hinting that the original painting might have shown a cluster of glyphs similar to those we saw in the *Codex Cruz-Badiano*. And on page 310 we see *Tuna*, or the nopal cactus, with its roots replaced by sandwiched stone glyphs that appear to be shown in profile (figure 9). When we compare these images to those of the *Codex Cruz-Badiano*, the consistency is striking.

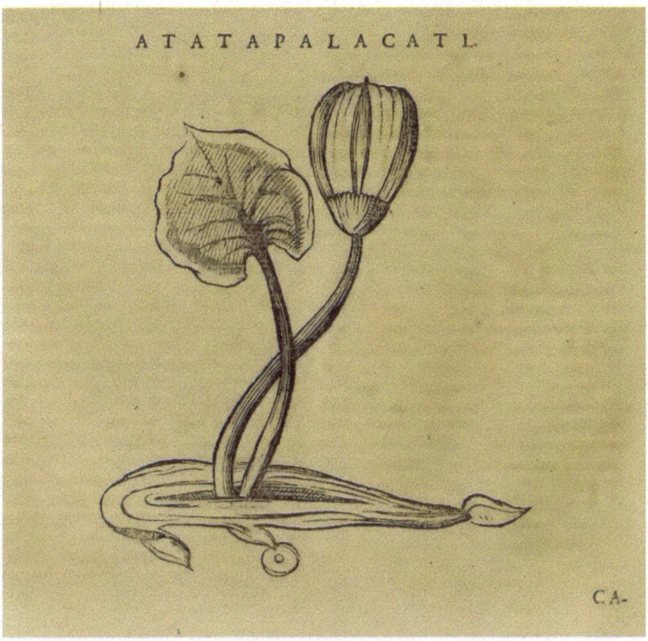

FIGURE 7 Atatapalacatl, from Juan Eusebio Nieremberg, *Historia naturae maxime peregrinae* (Antwerp, 1635), 306. Courtesy of Getty Research Library.

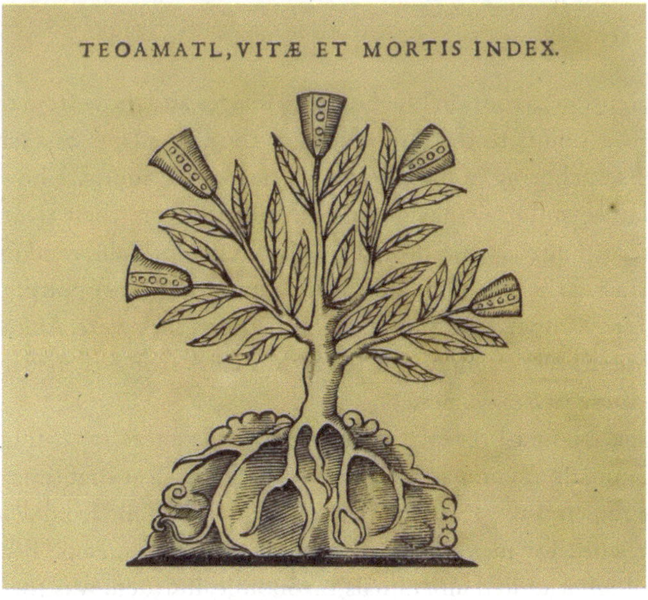

FIGURE 8 Teoamatl, from Juan Eusebio Nieremberg, *Historia naturae maxime peregrinae* (Antwerp, 1635), 308. Courtesy of Getty Research Library.

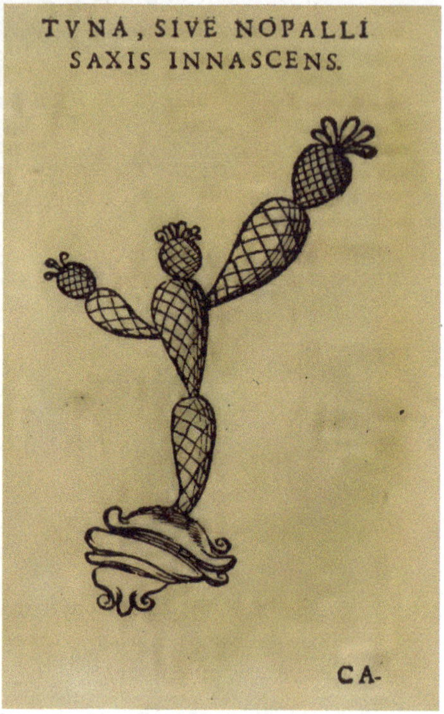

FIGURE 9 Tuna, from Juan Eusebio Nieremberg, *Historia naturae maxime peregrinae* (Antwerp, 1635), 310. Courtesy of Getty Research Library.

The attention to roots in Nieremberg's images and those from the *Codex Cruz-Badiano* points to the importance of showing plants as anchored in place. This anchoring in place becomes clear if we compare how Nahua nomenclature and drawings of animals differed from their treatment of plants. According to Bernardo Ortiz de Montellano, Nahua and European peoples shared the view that animals and minerals were opposite one another in terms of what kinds of natural creations they were. Animals were alive; minerals were simply of the earth. Plants stood at that midpoint between these two extremes.

Classical Náhuatl reveals some of the differences regarding Nahua views of animals as compared to plants. In Náhuatl, at that time, animals could be referred to generally as *yolqui*, which is built from *ollin*. From this root word for movement, we get *yóllotl*, which means "heart." The essence of what united all animals taxonomically, then, was their ability to engage in movement through the power supplied by their heart. Tellingly, Náhuatl at that time had no general term for plants; that is, there was no plant equivalent of yolqui.[45] If what held animals together was a heart that enabled autonomous movement across regions, the thing that united plants was that they were anchored in place, whether in rocky soil or marshy land, which was conveyed through the way the plant was illustrated.

Attention to how classical Náhuatl used plurals reveals another aspect of Nahua perceptions of different forms of nature. According to lexicographer Rémi Siméon, plural forms in classical Náhuatl were reserved for sentient beings and could not be applied to plants or minerals. To indicate more than one the noun is not modified as it is for animals. Instead, the adjective *miec* is placed before the noun, to mean "many." Thus, while *quauhtli* (eagle) can be pluralized as *quauhtin*, the word *xochitl* (flower) cannot be rendered as a plural. One would have had to say miec xochitl, or many flowers.[46]

European naturalists, like the native Nahua, understood plants and animals to have different relationships to place. We can see this in José de Acosta's *Historia natural y moral de las Indias* (*Natural and Moral History of the Indies*). The Spanish Jesuit Acosta (c. 1539–1600) lived in Mexico from 1586 to 1587. In 1590, after his return to Spain, he published his *Historia natural*, which was soon translated into other languages (including English in 1604) with broad circulation across Europe. Acosta divided nature into three branches—Metals, Plants, and Beasts—and attached to each a different relationship to place. Metals could be completely extracted from the soil, with no lingering attachment to place. Plants, however, were "living creatures fixed in one place."[47] For Acosta, animals and people happened to be in the New World incidentally, but plants were firmly rooted to a particular location.

For Nahua and Europeans alike, then, New World plants were fundamentally different from peripatetic animals and bore a more intimate relationship to place. Moreover, Acosta considered New World plants to be *more* firmly anchored to place than those from Europe. According to Acosta, this explained why Spanish trees and plants readily flourished in the New World, while New World trees and plants resisted acclimatization to Europe. This was paralleled by assumptions that European culture was expandable and appropriate for all places in the empire, whereas the indigenous cultures of the New World were only fit for that place, an assumption that Hernández, in his embrace of Náhuatl taxonomy, threatened to upend.

Returning to Nieremberg's reproduction of Hernández's images, we can see how the tlacuiloque similarly anchored the teoamatl plant to its particular place in the New World. Schematized flowers and leaves lead down to thicker stems and a lumpy, quirky main trunk. This trunk is firmly rooted in a mound of earth. The soil edges are rendered in glyphic form, with curls and curves that seem to suggest either a jumble of stone glyphs piled together to form a place glyph or an improvisation on a standard manner of depicting place-names. Many examples of this standard manner of depicting place-names can be found in other codices, including the *Codex Telleriano-Remensis*, c. 1653, in which the place glyphs for towns such as Bixatitan (figure 10) exhibit a use of curled stone-glyph tips similar to what we see in Nieremberg's teoamatl. Thus, the soil to which the roots of Nieremberg's teoamatl cling might have been, in Hernández's original, a knot of stone glyphs, a place glyph, or a creative merger of the two.

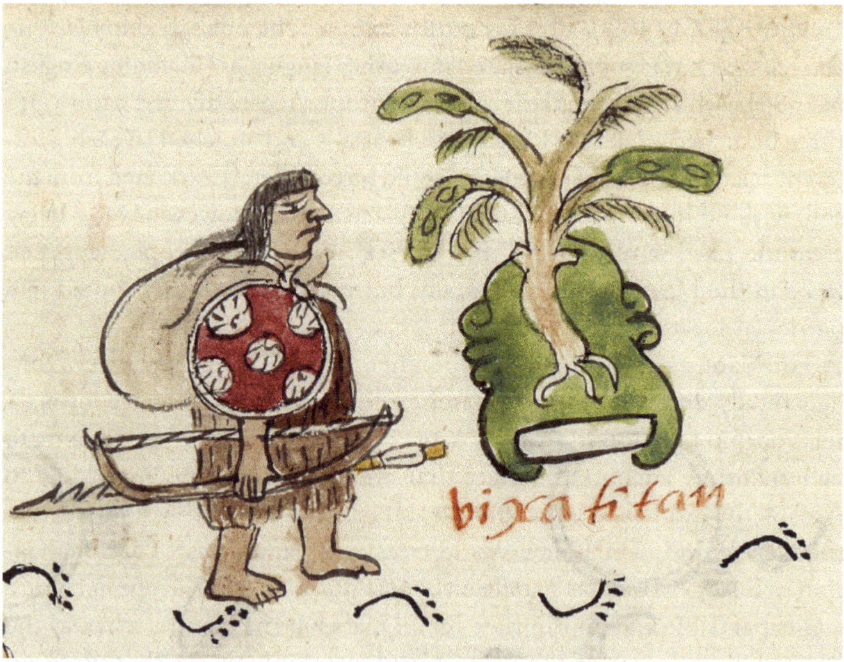

FIGURE 10 *Codex Telleriano-Remensis*, c. 1563, folio 282. Courtesy of the Bibliothèque National de France.

The nopal with its tuna (prickly pear) fruit is similarly intriguing in the depiction of its growth habit, its fruits and flowers, and its spines (rendered as a three-dimensional grid wrapping around the plant's flesh; figure 9). The glyphic representation of the plant's soil is pronounced. As with the atatapalacatl, here the tlacuilo completely replaces the root with a depiction of the substrate out of which the plant grows, represented as a collection of stone glyphs compressed together. The top stone glyph sits almost like a Seussian gloved hand with lifted fingertips. At the bottom is a glyph shaped like an upside-down crown, perhaps suggesting the shape of its roots. Two flattened layers, one thicker than the other, sit sandwiched between the handlike glyph and the crownlike bottom.

Images such as these, paired with those from the *Codex Cruz-Badiano*, remind us of the previously quoted insight by historian Stuart McManus that "Renaissance humanism was not one thing. Rather, it was a versatile toolbox of ideas, linguistic techniques, and scholarly practices that could be put to multiple ends by humanists trained in its methods."[48] Native draftsmen should be understood as contributors to this Renaissance practice as much as any of their European counterparts during an era that Daniela Bleichmar and Peter Mancall, building on the work of James Axtell, have described as an "age of mutual discovery."[49]

DEFENDING NATIVE IMAGES

Hernández commented about the occasional "mistakes of the artists" that he had them correct.[50] We do not know exactly what these "mistakes" were or what corrections he had them make, but these images, combined with the esteem he regularly expressed for his draftsmen's work, suggest that Hernández did not consider it a mistake for his tlacuiloque collaborators to employ native visual vocabularies. He described their work as "large figures on large paper, all true to life and representing all the parts and proportions with greater and fresher exactness than ever before."[51] Hernández's eagerness to vouch for the "exactness" of the tlacuiloque's work suggests that he held these images in high regard, and that he hoped that they, along with the Náhuatl nomenclature system, would play a role in his revision of Renaissance natural history.

His repeated emphasis on the novelty and exactness of his draftsmen's images might have been merely an effort to exaggerate their quality to gain favor with the king but, given the context of how he communicated such praise, this is unlikely. Hernández was very open about when he lacked confidence in particular guides, assistants, or local officials. And when he wanted to curry favor with the king, he did so by immodestly touting his own abilities or reminding the sovereign about the transformative implications of his mission, not by unduly exaggerating the skills of indigenous people.[52] As did most botanists of the Renaissance through the Enlightenment, Hernández wrote of the need to keep a close eye on his draftsmen, to make certain that they consistently drew each plant "as it was in reality" and according to his instructions.[53] But he expressed unwavering confidence in the quality and accuracy of the images he commissioned and sent to the king.

There is evidence that he might have worried whether other Europeans would share his opinion. We see this, for example, in his paranoia about how his patron might handle the images in his absence. In 1572 and in 1576, he sent the king batches of plant illustrations by the native draftsmen, along with text, but intentionally did not correlate the images with the text. He explained to his king that he wanted to avoid correlating the illustrations and the text until the final moment before the work was printed so as "to prevent damage to the paintings from emendations." Hernández did not explain what kinds of "emendations," that is, what kind of revisions or corrections, he feared, but evidence suggests that he was worried that an editor might remove the indigenous elements that he seems to have most valued about the work of his tlacuiloque.

To further dissuade anyone from trying to mess with his images, he cautioned that he had intentionally placed them out of order, and even included duplicates which others might not recognize as such. Some plants, he explained, "are depicted twice or more, either as a precaution, or because they were painted in different regions and at different stages; but it will be easy to choose the most appropriate one at the time of printing."[54] The possibility that Hernández was trying to prevent interference with his images is made even more likely by a petition he wrote to the king in 1577 as he prepared to depart from Mexico. Perhaps worried that others might carry out emendation to his images before he could arrive in Spain to stop them, he urged the king not to allow anyone to begin work on the materials he had sent. He argued that no one else will "know how to put it in the right order and make proper use of it, and thus the nation will be disadvantaged. It will not be possible to make good this loss however many years go by, because a great number of doctors and Indian artists who contributed to this and made sense of it have died in the latest plague."[55] Thus, if someone damaged or altered the images or undid his structure, it would not be possible to fix it because the native experts were no longer alive. It turns out that his fears were justified, as Recchi and Cesi later proved.

Not once in his many letters to the king or in the surviving accounts of his experiences does Hernández express any desire to replace native artists with Spanish painters, or to have European artists redo the illustrations created by the tlacuiloque. Far from seeking artists who might be free of native practices, it is precisely the native practices of illustration, mediated by his own instructions, that Hernández seems to have valued and that he placed at the center of his intellectual project alongside his version of the Náhuatl taxonomic system. The nomenclature and visual images anchored the plants to their particular places and cultures of origin, without inhibiting their potential for acclimatization to new places and new purposes.

CONCLUSION

After he began working with native titicih and tlacuiloque, Hernández asked his king to order that he be sent descriptions and sketches of "everything natural" found in the Canary Islands, China, and other parts of the Americas "with an account of their virtues and qualities, supplied by native and Spanish doctors." He explained that, upon receipt of this material, he would assign them names and have each of these plants "painted over here the way I want it, in my style and manner, as no one else could initiate or complete this. Your Majesty will have a model, which the world will admire."[56] His request left little doubt that he expected to transform European science not just by discovering thousands of new plants, but also by introducing a new system of taxonomy and botanical illustration based on what he had learned from his Nahua collaborators. Hernández seems to have hoped that these systems could be used to anchor *all* plants, from across the globe, to their particular places and contexts of origin in ways that would enhance, rather than limit, scientists' and empires' ability to identify, possess, transfer, and exploit them at a level without precedent.

In the process, Hernández situated himself as the indispensable mediator between this Mexican system and its application first to New World plants and then to global nature. When his manuscripts and illustrations arrived in Spain, they were warmly received, and the Council of the Indies urged that they "should be published at your Majesty's expense."[57]

This initial enthusiasm abruptly soured in 1580, when Hernández was removed from his post. By fall 1582, the Crown was taking sweeping steps against the perpetuation of indigenous knowledge. Historical anthropologist George Baudot has shown that royal officials became concerned that texts and initiatives that validated indigenous culture might foment "a collective pride that could lead to" rebellion, "especially because the Indians kept trying in a thousand ways to return to their original mores. For the Council of the Indies, the Indian was still a problem."[58] In a letter dated September 28, 1582, King Philip II informed members of the Council of the Indies of his change of heart: "I have decided . . . that the books on herbs and animals and other things that Doctor Francisco Hernández wrote, and I believe are in the Council, be turned over completely to Bartolomé de San Toyo so that he may put them where I tell him." San Toyo was keeper of the king's jewels and royal secretary. While it is possible that the manuscripts were handed off to him simply for safekeeping, Baudot's argument suggests that it more likely was to restrict their circulation. This possibility is strengthened when we take into account that the Crown at this time prohibited the initiation of any further projects of the sort, and even censored the work of Sahagún.[59] Dora Weiner suggests that part of the reason why King Philip II "kept his *protomédicato* in exile and censored publication of the *Natural History*" might have been that "Spanish theologians perceived Hernández's naming of animals and plants as a heretical activity, because only God has the power to name and thus create living beings."[60] In the end, the king's change of mind regarding Hernández's texts might have been partly influenced by concern about the protomédico's hubris, but the stronger motivation and main effect was the suppression of Nahua structures of knowledge and the elimination of the possibility of intellectually Mexicanizing European science.

Historians David Carrasco and Scott Sessions argue that tlacuiloque in preconquest Aztec society saw the images they created with ink and paint as suffused with divine meaning. Writing "with pictures was not only a way of transmitting information and knowledge but also . . . the deepest truths of the community and therefore God and the gods, into the viewers' and listeners' minds."[61] While it is unclear whether this religious sensibility carried into these postconquest botanical illustrations, it is evident that Hernández and his draftsmen believed their images could reveal truths about nature that went beyond direct observation of the plant, and beyond the alphabetic text. Hernández embraced these artists' methods, including their improvisations, to meet the new needs of the Renaissance era.

European scientists have a well-earned reputation for their blindness to what other cultures have had to offer regarding knowledge about the natural world, but Hernández's experience in Mexico should prompt us to be cautious about presuming the inevitability of such a stance. It suggests a moment when a fertile cultural exchange was still imaginable. Hernández's work, albeit in distorted form, would be revisited and debated time and again in the coming centuries in relation to changing perceptions of science, indigeneity, and imperial-cum-national identity. This was aided by Hernández's insistence throughout his time in Mexico on carefully distinguishing "between those [plants] that are native [to New Spain] and those that have been transplanted."[62]

This attention to place would prove particularly compelling to future generations of Mexicans eager to anchor themselves to their natural world. In the nineteenth century, nationalists who were searching for an autochthonous identity for their struggling republic, and who were eager to claim the products of nature found within its national territory, would come to embrace Hernández and his findings.

When the Mexican Pavilion at the 1893 Chicago world's fair displayed the previously lost manuscripts from the Royal Botanical Expedition to New Spain (1787–1803), as mentioned in the introduction and discussed in chapter 7, it demonstrated that claims to cosmopolitanism could go hand-in-hand with assertions of patriotism and self-determination. Mexican scientific intellectuals and state leaders used this global stage to proclaim their ability, as well as their right, to control the natural resources within their territory. This right was asserted against foreign interests, whether from Britain, Spain, or the United States, and it depended on the presumption that the modern republic was the rightful heir to the ancient indigenous civilizations, including their knowledge of and claims to the natural world. It also left no doubt that Mexico, rather than Spain, was the rightful inheritor of the colonial-era scientific discoveries that had been carried out within Mexican territory. To Mexican scientific intellectuals at the time, the Renaissance-era Hernández Expedition, the Enlightenment-era Royal Botanical Expedition, and the work of late nineteenth-century Mexican scientific societies were links in a single chain. And so, it is to the second link of this chain that we now turn in the next chapter: the late eighteenth-century Royal Botanical Expedition to New Spain.

PART II

THE ROYAL BOTANICAL EXPEDITION TO NEW SPAIN, 1787–1803

CHAPTER 3

NATURE AND KNOWLEDGE IN ENLIGHTENED NEW SPAIN

Vicente Cervantes, a member of the Royal Botanical Expedition to New Spain (1787–1803), claimed to be indifferent about the names people used for particular plants when talking in the "plazas or in small gatherings with Indian herbalists and vegetable vendors," but he judged those names to be inappropriate for "places of learning."[1] He did not necessarily seek to blot out popular modes of knowing nature, but he did trivialize them as unworthy of modern science. This trivialization of native knowledge was a sharp contrast to the Hernández expedition from two centuries earlier, which the Royal Botanical Expedition to New Spain was designed to succeed.

The Crown hoped that its Royal Botanical Expedition might help Spain reassert its imperial domination over its colonies, proclaim its relevance to the Enlightenment, and better equip it to challenge its European rivals. This ambitious expedition has inspired a historical literature of impressive volume and quality. Most of these studies examine the documents produced by the expedition to help untangle species identifications. They also offer insightful analysis of the undertaking itself, to trace its contributions to the professionalization of biology and chemistry, reconstruct the members' travel itineraries, narrate bureaucratic and personality conflicts, and enhance our understanding of the Bourbon Reforms.[2] The chapters in this section build on this impressive literature, but they approach the expedition from the perspective of environmental history, to ask what the expedition's activities within New Spain can teach us about how changing understandings of nature contributed to a Mexican imagined community and reinforced social inequalities. In this sense, these chapters expand on, while also revising, insights by scholars who have described the rise of New World creole protonationalism during the late eighteenth and early nineteenth centuries.[3]

This section also argues that expedition members and their supporters did not attempt to obliterate the importance of place, which scholars sometimes claim was a goal of Enlightenment scientists. On the contrary, they asserted Spain's imperial power by anchoring plants specifically to the viceroyalty of New Spain, with Mexico City as the site for concentrating knowledge and resources for transmission to the metropole. Because of this, even as they sought to configure understandings of space to meet the evolving needs of the Spanish Crown's absolutist imperial ambitions, members of the expedition and their supporters unwittingly created new possibilities to imagine alternatives to the very imperial structure the expedition was designed to enforce.

Like Hernández two centuries earlier, members of the expedition, along with the generations of Mexican scientists they eventually trained, emphasized the importance of place. But they did so in a very different manner than had Hernández, and with different implications. They worked under the presumption that the study and organization of nature needed to be removed from the hands of indigenous people and anyone else not trained in particular scientific methods emanating from the metropole. In their view, Enlightenment men of science were exclusively qualified to oversee a hierarchical structure of knowledge about the natural world. They deemed other modes of knowledge and organization inferior and folkloric, especially those derived from indigenous practices. Useful knowledge appropriated from non-European sources, meantime, was recast as though it were the product of European genius, erasing as much as possible its indigenous roots.

From their base in Mexico City, the members of the expedition codified their view of the natural world as the new common sense, in the Gramscian sense of ideas and power relations that become normalized almost to the point of erasing the genealogies of their own creation.[4] This new common sense guided how scientists, the state, and economic and political elites interacted with Mexican nature during the late-colonial era. The expedition's linkages between plants and place, along with the hierarchy of knowledge it imposed, would later become useful to nineteenth-century nationalists eager to assert the sovereignty of the Mexican Republic, while also empowering the government and the country's elite to lay claim to the land and natural resources of the former viceroyalty that it now defined as its own patrimony. The chapters of this section trace how the Royal Botanical Expedition transformed understandings of the natural world within New Spain, while paying particular attention to the ways its members deployed visual images and the organization of physical space in pursuit of their endeavors.

RECAPTURING IMPERIAL GLORY: ENLIGHTENMENT SCIENCE AND NEW WORLD NATURE

The Royal Botanical Expedition to New Spain was part of a broader strategy by the government of King Charles III during the late 1700s to modernize its vast holdings into an efficient extractive empire along the lines of its English, Dutch, and French rivals, while also identifying new untapped sources of wealth. New Spain's silver mines had made the colony an economic powerhouse for the Spanish Empire. But this silver was passing straight out of Spanish hands, creating only fleeting wealth for the Crown. It was hoped that botany might provide a more stable foundation for long-term prosperity. As one Spanish official revealingly declared, "a dozen naturalists and some chemists scattered in Spain's dominions . . . will offer an incomparably larger utility to the state than a hundred thousand men fighting for the enlargement of the Spanish empire."[5]

As Spanish Enlightenment-era scholars anxiously debated the past, present, and future of their empire, they latched on to Francisco Hernández's expedition of 1570–1577 as a point of historical pride as well as a source of information to guide future efforts.[6] And they at last began to appreciate the magnitude of the loss in 1671, when a fire had torn through the royal palace of El Escorial and consumed all Hernández's manuscripts and illustrations.

Eager to recover any fragments of Hernández's work that might have survived in Mexico, Casimiro Gómez Ortega (1741–1818), head of the

Royal Botanical Garden in Madrid, hired three intellectuals in New Spain to comb through the Mexican archives. His first two agents were the well-known Creoles José Antonio de Alzate y Ramírez (1738–1799), who since the age of eighteen had been obsessively scouring the archives for such texts with no success, and José Ignacio Bartolache (1739–1790), who seems to have indifferently expended no energy on the task. Gómez's third agent, who became the central figure in the Royal Botanical Expedition, was Martín Sessé y Lacasta (1751–1808, sometimes signed as Cecé), who originated from Zaragoza, Spain. Possessed of medical training, a talent for scientific research, and a taste for adventure, Sessé had served as a military doctor in Gibraltar and Cuba, where he conducted clinical trials to combat the epidemics that were debilitating the imperial army. In 1785, he moved to Mexico City to collect on a 1,000 peso loan he had extended to the Crown. There, he took over a lucrative private medical practice and worked at research hospitals.[7] After the trio failed to find any shreds of Hernández's original manuscripts, Sessé, ever the optimist, convinced the disappointed Gómez Ortega to turn this setback into an opportunity.

Gómez Ortega was a towering figure in European science who used his position as head of the Royal Botanical Garden in Madrid to fiercely advocate for the Linnaean system in botany and its chemistry counterpart, the Lavoisierian system. He presents an instructive contrast to another towering botanical figure of that time, Joseph Banks (1743–1820) of England, who helped shape the Royal Botanical Gardens in Kew, London. Though both men put science at the service of empire, Banks was not trained as a scientist, and his interest in plants was part of his broader fascination with curiosities and things exotic.[8] Gómez Ortega, by contrast, was a scientist, rigorously trained in the latest botanical and chemistry methods. When Sessé in 1785 proposed a new expedition, Gómez Ortega embraced the idea. They quickly won the support of both the New Spanish viceroy and the Intendant of the Indies, and then presented their plan directly to the Crown, which in 1787 offered royal patronage.

The initial justification for the expedition was to recreate Hernández's lost illustrations and to fact-check and update his text (insofar as his text was understood through the distorted interpretations by Nardo Antonio Recchi and Federico Cesi; see chapter 2).[9] As was stated in the Royal Order of March 20, 1787, with which King Charles III authorized the expedition:

> [I order that] the natural products of my fertile domains of New Spain be methodically examined, drawn and described, not only with the general and important object of promoting the progress of the physical sciences . . . but also with the special [objective] of supplementing, illustrating and perfecting, in accordance with the current state of the same sciences . . . the writings left by Dr. Francisco Hernández [which were] financed by the Monarchy but until now have not produced the full profits that should have been expected from it.[10]

Gómez Ortega and Sessé intended to comply with the Crown's order, but they also were hatching larger plans, including the creation of a botanical garden and a school of botany and medicine in Mexico City, for which they soon also won royal support.[11] They envisioned that the research expedition, the garden, and the school would work in tandem to conduct original research, reform the medical-training-and-licensing establishment in New Spain, and free the colony and the mother country from reliance on medicinal imports from imperial rivals—medicines that were crucial for Spanish military troops.[12]

Sessé and Gómez Ortega shared a desire to impose a top-down structure in which the metropole was to serve as the hub for the promulgation of scientific uniformity across the empire. Recently launched colonial expeditions to other parts of its empire, such as those by Hipólito Ruiz and José Antonio Pavón to Peru and Chile (1777–1788), José Celestino Mútis to Nueva Granada (1783–1808), and Juan de Cuéllar to the Philippines (1786–1797), had been initiated in the absence of central coordination and without any plan about what to do with the information they gathered. Gómez Ortega created the Royal Office of Botany (Oficina Botánica) to lend order to these existing expeditions, while designing from the ground up the Royal Botanical Expedition to New Spain. This newest expedition would be guided by the latest scientific theories and methods to bolster Spain's scientific reputation and rationalize the Crown's exploitation of New World nature.[13]

By the middle of 1787, Sessé and Gómez Ortega had recruited much of their expeditionary team. Most importantly they had hired Vicente Cervantes (c. 1755–1829) as head of the garden and *catedrático* (chaired professor or director) of the school of botany and medicine. After entering an apprenticeship in a Madrid drug dispensary as a youth, Cervantes had pursued his intellectual interest in botany. Despite being an autodidact, he brilliantly passed the medical certification exam. This rare feat attracted the attention of Gómez Ortega, who helped him win a professorship at Madrid's Royal School of Botany and a position as botanist to the General Hospital of Madrid.[14] Once the Crown had endorsed the Royal Botanical Expedition to New Spain, Gómez Ortega appointed his rising protégé to assist Sessé in the creation of a Mexican garden and school of botany and medicine, which were to follow the models in Madrid. Cervantes arrived in Mexico in 1787 to begin his collaboration with Sessé and the other members of the team.[15]

Between 1787 and 1803, the ten or so members of the expedition traversed thousands of miles, usually divided into two or more teams at a time to cover more ground. While most of their efforts focused on areas that would become modern-day Mexico, they also traveled south through Guatemala into Panama, east into Venezuela and Barlovento (modern-day Lesser Antilles of the Caribbean), and north through California to Nutka (an ill-defined region roughly equivalent to the southern section of modern-day British Columbia, Canada, and portions of the northwest of the modern-day United States).[16] (See map 1.) They collected seeds, live plants, and dried herbarium samples, and described and drew thousands of plants. The expedition lasted almost seventeen years, making it among the longest of its era. During these years, it also reorganized botanical education and medical practice in New Spain, and created Mexico's first Enlightenment-influenced botanical garden and its first natural history museum.

Members of the expedition set themselves against what they viewed as the backward practices of the local medical establishment. In this conflict, members of the expedition benefited from the weight of the Crown. Whereas the king had forbidden the earlier Hernández expedition from interfering with local licensing or medical practices, Sessé won a royal mandate to meddle *precisely* in these matters.[17] He ignited tension at the start by ordering the conversion of several busy general hospitals into research facilities, with fewer beds. When the Mexican medical board balked, the Crown overruled it to grant Sessé the control he sought.[18] This was the start of a tempestuous relationship in which members of the expedition insisted on their right to operate unhindered by local officials, while members of the university and medical establishment protected their own interests by erecting endless bureaucratic hurdles for Sessé and his team.

The expedition also came into conflict with local naturalists unaffiliated with the official medical establishment, but who operated outside the Linnaean method, most notably the Creole Enlightenment intellectual José Antonio de Alzate. In 1788, just as the expedition was trying to establish itself within Mexico City, Alzate launched his *Gaceta de literatura*. The journal was structured around the argument that foreigners lacked the epistemological knowledge to comprehend the human and natural world of the Americas. Instead of relying on metropolitan expertise, Alzate commissioned reports from local priests across Mexico with deep knowledge of local cultures and environments. And, like his intellectual predecessor Francisco Javier Clavijero (1731–1787), Alzate embraced Hernández's work, which had been built partly on Nahua science. He even used the journal to directly refute the value of the Linnaean system, particularly its separation of the signifier from any characteristics of the object being signified. In these ways Alzate rejected the authority of the European-based Enlightenment in favor of his own home-grown Mexican movement.[19] Historian Dorothy Tanck de Estrada shows that the expedition became part of the broader struggle between Criollos (Creoles, those of European extraction born in the Americas) and *Peninsulares* (Peninsulars, those born in Europe) that unfolded during the Bourbon Reforms.[20] While her point is well taken, that Creoles soon comprised half the membership of the expedition at all ranks complicates any effort to reduce these conflicts to a Peninsular–Criollo divide. For the expedition, the conflict revolved less around the birthplace of individuals than around hierarchies of structures of knowledge and whether systems of knowledge emanating from the metropole could be imposed upon local society.

As Cervantes's statement at the start of this chapter suggests, in addition to setting themselves against the entrenched local medical establishment and representatives of the homegrown Mexican Enlightenment, Sessé and his team also rejected the expertise of indigenous and mestizo herbalists. Lest we accept the expedition members' claim (which has been taken for granted by their later biographers) that they were stamping out backward traditions in favor of medical practices rooted in rationality and verifiable facts, we should keep in mind that many of the medical methods that they described as cutting-edge were still firmly rooted in humoric theory and other medical assumptions that, today, we might view as less advanced than the pharmacopeial approach of the native herbalists they publicly scorned. Hernández had relied openly on indigenous knowledge and even applauded the Nahuas' approach to naming plants. This new expedition, however, claimed to depart in crucial ways from its sixteenth-century predecessor by trying to steer clear of such tendencies.

Though the expedition was inspired by Hernandez's sixteenth-century efforts, its members had their eyes firmly fixed on the future, not on the past. Sessé described himself and his team as living in a new scientific era that would build on previous knowledge while rendering that past superfluous. He noted that "no sooner was America discovered and incorporated into the vast territory of the Crown of Castile than Philip I sent to this continent his illustrious physician Dr. don Francisco Hernández to identify, discover, and collect its medicinal products," which, he claimed, the Crown immediately recognized as more valuable even than precious metals. "Hernández," Sessé argued, "did his best to fulfill the good intentions of the monarch, but the lack of scientific principles, which were not known until 200 years after his death, made it such that his descriptions, despite being no worse than those of the wisest Greeks and Romans who preceded him, and superior to those of almost any other botanist since Pliny, turned out to be so incomplete and defective that" they could offer no real guidance to modern science. He credited Hernández for having extracted, tested, and refined indigenous knowledge, and admitted that the native informants "certainly knew more useful plants than those described by Dioscorides." Had Hernández's findings been fully exploited in Europe in the 1500s, they would have added greatly "to the riches of the old continent." But that did not happen, and according to Sessé it now was too late. The tools provided by Enlightenment science were so powerful, in his view, that they eliminated the need to mine native knowledge; any effort in that direction would bring natural history up to date with the sixteenth century, but were of no use, he argued, to the future of modern science.[21]

He claimed that the expedition's Enlightenment methods offered unprecedented precision. This enabled it to accomplish "the feat of collecting, identifying, and discovering more than 3,000 species, illustrating most of these with great exactitude, and objectively describing their distinctive characteristics so that for all time and in all places they can be recognized without the slightest doubt." According to Sessé, modern theory and scientific methods of experimentation rendered local knowledge and traditions (including locally developed forms of creole science) superfluous. And his expedition's mastery and deployment of these methods even made it superior to all the other recent expeditions that had been carried out by Spain—including those to Peru, Chile, Nueva Granada, and the Philippines.[22]

IMPERIAL DOMINATION AND THE SEEDS OF NEW SPAIN

To the frustration of later botanists, the expedition's scientists usually did not record the particular location where they collected or drew a specimen, nor did they comment on the plant's range or habitat. All that they conveyed, all that seemed to matter, was that the plants were from the Crown's colony of New Spain. By abstracting plants from their webs of local context, they reconfigured space and knowledge, centralizing it within Mexico City. By implanting metropolitan institutions and approaches within the colony and training several generations of Mexican doctors, scientists, and pharmacists in these new methods, the expedition changed how nature was viewed within New Spain, and how it was understood in relation to the imagined community and to the territory dominated by Mexico City.

Members of the expedition were unambiguous about privileging metropolitan structures of knowledge over all other ways of knowing. This was as true for the Spanish heads of the expedition, Cervantes and Sessé, as much as for local Creole recruits, including José Mariano Mociño (1757–1820, sometimes signed as Moziño), Sessé's Creole protégé and eventual successor from Temascaltepec, who had been trained in the expedition's school of botany and medicine and later became a leading figure in Madrid's scientific academy. But their desire to banish indigenous knowledge from the halls of science could not be as thorough as they led their audiences to believe, because Enlightenment science ultimately depended on the forms of knowledge that it belittled and sought to displace.

Back during the Renaissance, when a naturalist discussed a particular simple, he attempted to ensure that he and his audience were thinking of the same plant by listing its many vernacular and Latinate names, and he perhaps also described the various stages of its growth.[23] During the

Enlightenment, the system devised by the Swedish botanist Carl Linnaeus was bringing order to the naming and categorization of plants, but, if a scientist wanted to be certain that he and his audience were talking about the same plant, it was not enough to simply state the Linnaean Latinate binomial: he still needed to list the many vernacular and scientific names the plant had acquired over time. This was true even in the teaching methods of the expedition's school of botany and medicine, where, to train his students to make practical use of their education, Cervantes had to move outside the Linnaean system by continuing to list plants' local vernacular names. Thus, even though members of the expedition did not endorse the native nomenclature system that had fascinated Hernández, they could not dispense with Náhuatl names. Rather than blot out other modes of knowing, the expedition had to settle for imposing a hierarchy of knowledge that exalted and masculinized the Linnaean system, reinforced the authority of European centers of science, and infantilized other ways of organizing knowledge.

Ironically, even as the expedition subordinated the colony to the metropole and created a hierarchy of knowledge, it also nurtured possibilities for conceiving of New Spain as a distinct imagined community rooted in nature and centered in Mexico City. By venturing across the breadth and width of New Spain, far more broadly than even Hernández had done, this expedition was the first effort to define the colony not by the political structures that held it together, by the peoples it ruled over, or even by its property lines, but by the plants that buried their roots in its soil.

THE ROYAL BOTANICAL GARDEN IN MEXICO

The lynchpin of the expedition was its garden, because it was into this garden's beds that the scientists planned to transplant all the species they collected from across New Spain. And it was the garden (as an institution) that was responsible for running the school of botany and medicine, which would implant Enlightenment scientific methods within the colony. Such was its importance that, before Sessé was willing to set off to explore the countryside, he wanted to resolve the many procedural hurdles erected by the local functionaries and rivals who tried to thwart its inauguration.[24]

After overcoming some of these bureaucratic obstacles, Sessé and Cervantes officially inaugurated the Royal Botanical Garden on May 1, 1788. The ceremony included the regent of the Audiencia, a representative of the viceroy, attendees from the university, and members of Mexico's scientific societies, along with other distinguished persons of the capital. After an orchestral performance, Sessé delivered a lecture on innovations in the study of botany and medicine. Then followed a firework display by the most renowned pyrotechnician in the colony, Joaquín Gavilán, that incorporated colorful explosions in the form of recently discovered Mexican plants, and, reportedly, even a male papaya tree shooting flares of pollen to fertilize two nearby female papayas.[25] This attention to flower structures and pollination emphasized the primacy of the Linnaean system, which categorized plants on the basis of their reproductive features. It also spoke to the ambition of the expedition to spread, like the male plant, its pollen of scientific knowledge to fertilize the feminized colonies so that they, through the rationalized exploitation of their natural resources, could provide nourishment to the empire.

During the event, the rector of the university was required to affirm the investiture of Sessé and Cervantes as his equals. Sessé and Cervantes already had found themselves at odds with the members of the university, so this ceremonial display by the rector was of no small significance. It eradicated jurisdictional claims by members of the university who had tried to exert supervision over the expedition, and it publicly affirmed the expedition's right to have a say in local practices. The event coincided with royal orders requiring that all medical students at the university complete courses in the garden's school of botany and medicine, and that the head of the garden was even in charge of administering the medical licensing examination. The next day saw the inaugural class session in botany, filled with aspiring doctors eager to learn European science and, above all, to understand the practices and theories that had developed around the illustrious Linnaeus.[26] This first crop of students included José Mariano Mociño, the star pupil who later joined the project in 1790 and eventually ascended to leadership of the expedition, and José Maldonado (dates unknown), who also was hired by the expedition.

The garden's inauguration marked a triumph for Sessé. Along with the Seminario de Minería (School of Mining) and the Academia de San Carlos de Arte (San Carlos Academy of Art), the garden—including its school of botany and medicine—was one of the three academic institutions created as part of the Bourbon Reforms to promote intellectual growth and economic prosperity in New Spain while reinforcing regal authority.[27] The problem was that, despite the formal inauguration of the Royal Botanical Garden in 1788 and all the fanfare, there was not yet a real garden.

At the time of its inauguration, it was still more of a concept than a reality. The vision for how the Royal Botanical Garden in Mexico City would operate was first described by Viceroy Bernardo de Gálvez in August 1785, right after he and Gómez Ortega had become captivated by Sessé's proposal for the expedition. Viceroy Gálvez envisioned a Mexican garden laid out and operated "following the idea and in imitation of the manner of the Royal Botanical Garden in Madrid, which is the expected source for the methods, rules and teachings." Though it would be designed to look like a copy of the Madrid garden, it was not to be on equal footing, nor did it serve the same purpose. As the viceroy explained, a central mission of the Mexican garden would be to operate as a "valuable repository for the products of North America so that they might be moved to the Metropole."[28] Sessé similarly emphasized the garden as a tool of empire. While making a case to the king for why the garden should be near the heart of the capital, he wrote: "Ultimately the benefit of being in the middle of the City, on a principal avenue, in imitation of your own court, beyond drawing attention to the unity of this Empire, would be a sweet attraction for all People of good taste, and will excite the admiration of many passersby to take part in scientific learning."[29]

The plan that ultimately emerged was for a garden that would imitate two very different things. First, it would imitate the Madrid garden to import European Enlightenment science into the colony, and in the process assert New Spain's subordination to the metropole. Second, there was the desire for the garden to replicate the botanical diversity of New Spain, thereby asserting Mexico City's control over the resources of this vast region.

In her study of the role of botany in British imperialism, Beth Fowkes Tobin explains "the Palm House [in England's Kew Gardens] represents the triumph of Britain's power to collect, analyze, and manage the world's resources for Britain's benefit."[30] In New Spain, we see a similar process unfold in the Mexican botanical garden, which was a celebration of the power of empire and Enlightenment science to collect, analyze, and manage resources. But unlike Kew or Madrid this garden did not fixate on the idea of "the tropics," nor did it aspire to global representation. Rather, it sought to collect, analyze, and manage those resources bounded by the spatial claims of the colony of New Spain. Its goal was to study and organize the products from this territory for exploitation by the empire. From its first conception, therefore, the garden would not just reinforce the empire, but also the integrity of nature within New Spain, centered in Mexico City—a spatial relationship whose impact would persist long after the fall of the Spanish Empire.

The Crown issued remarkably specific instructions about how the garden was to put these ideals into practice. These instructions, likely drafted for the Crown by Gómez Ortega, first clarified beyond any doubt that the garden would operate under the immediate protection of the king and his representative the viceroy, with resource requests handled by the Secretaría del Despacho Universal de Indias, a Bourbon institution that answered directly to the king, bypassing local bureaucracy. This removed the garden from the oversight of local authorities.[31]

The royal orders then instructed the expedition's researchers to transplant all the plants they collected into the garden. The garden had to keep a running log of everything that moved through its beds and coordinate the transfer of these items to the "Royal Botanical Garden in Madrid," which would work together "with the other Gardens of Spain and other major ones of Europe" to further study these plants. This made clear precisely how the Mexican garden was to relate to the other gardens in the international circuit, as well as how natural products were to flow within that circuit (i.e., through Madrid).

The instructions even clarified how the garden would enlighten the people of Mexico. Members of the clergy, military, and government, along with other "well-dressed urban individuals" would be permitted to stroll through its grounds, while provisions should be made so that the general public could clearly view the garden beds from a distance. In this way, people from all walks of life in the colony could bear witness to the enlightening benefits of this imperial institution. The orders also detailed exactly how the head of the garden was to administer the local medical licensing examinations. And it ordered that garden employees were required to hone their skills of scientific observations by completing botanical illustration classes at the Academia de San Carlos. The instruction also specified that the garden would continue its work in Mexico even after the rest of the expedition eventually completed its term.

To fulfill the various aspirations, Sessé sought a space for the garden and its school of botany and medicine near the viceregal palace and university, the centers of learning and political power in the heart of Mexico City. When his request for this venue became inextricably tangled in bureaucratic red tape, he turned to a zone known as the Potrero de Atlampa (Atlampa Pasture), halfway between the central square (Plaza Mayor, today known as the Zócalo) and Chapultepec. Sessé argued that the large plot "could not be better suited to the plan." It was ideally located along the Paseo de Bucareli, extended east to the Acequia de Belén (modern-day Balderas), north to the Paseo Nuevo (modern-day Avenida Morelos) and south to the Calzada de Chapultepec (along which ran the recently rebuilt aqueduct that carried water from Chapultepec hill to the new Salto de Agua fountain) (see map 3). At the proposed site's northeastern corner sat a house belonging to the prominent architect and *desagüe* engineer Ignacio Castera (1750–1811), who agreed to rent it for use as a classroom space.[32] The location also offered easy access to the Real Hospital de Indios and the Real Hospital de San Andrés, where Sessé planned to conduct clinical trials on indigenous and Spanish patients, respectively, using simples derived from plants grown in the garden.[33] By the time of the inauguration, Mexico City had transferred the Potrero de Atlampa to the Crown, but no work had begun at the site.

Once the formal inauguration was complete and the royal orders had been confirmed by the viceroy, Sessé and Cervantes were ready to start transforming the Potrero de Atlampa into a garden. The soil was of poor quality, but that did not worry Sessé, who was confident that modern science would bend nature to his will.[34]

Their plans immediately were delayed by a lawsuit claiming that a section of the Potrero de Atlampa was part of a Parcialidad de Indios. The indigenous litigants occupied the land in protest, demanding a rental payment, purchase at a fair price, or exchange for another lot of equal value.[35]

After the ownership issue was resolved—the documents are silent regarding how it was resolved—Sessé and Cervantes hired Miguel Constanzó (1741–1814, sometimes signed as Constansó), one of New Spain's leading architectural engineers, to design the garden and its structures, including a space for the school of botany and medicine.[36] Constanzó had

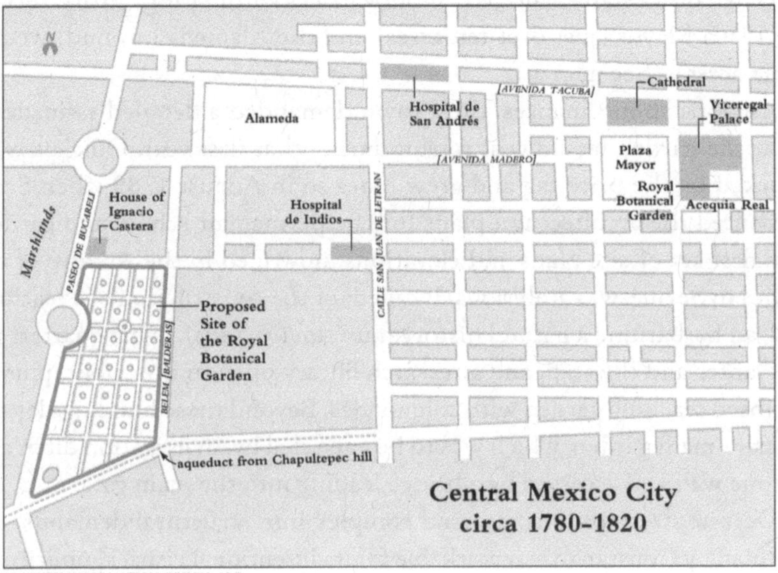

LUCIDITY INFORMATION DESIGN, LLC

MAP 3 Central Mexico City, circa 1780–1820.

arrived in New Spain in 1764 at the age of twenty-six as a member of the Corps of Engineers, then, in 1776, married into a prominent Mexican family. He took part in multiple expeditions to the northern regions, including California where he mapped San Francisco Bay and designed defensive structures, and has been hailed as California's "first true historian." He then returned to Mexico City. As one of New Spain's few formally trained engineers, Constanzó discovered that his skills were in great demand. He engaged in numerous high-profile projects that ranged from military fortifications to civic structures, road construction, and hydrological projects (including the desagüe project that aspired to build drainage tunnels leading out of the high valley to solve the city's flooding problems).[37] He seemed the ideal choice by the Royal Botanical Expedition for a project of such ambition.

At around fifty-six acres, Mexico's Royal Botanical Garden would have dwarfed its European counterparts, including Kew (which today stands at over one hundred acres but, at that time, was only eleven acres) and the Royal Botanical Garden in Madrid (about twenty-four acres). It also exceeded other state-managed outdoor spaces within Mexico City, such as the Plaza Mayor (just over ten acres) and the Alameda (around twenty-three acres). (See map 3.)

Anxious about finances, the Crown demanded a detailed estimate of what the garden would cost to run during that first year. Officials were shocked by the price tag and grew more so in August 1789 when Sessé presented the architectural plans for the permanent school and garden. The first set of site plans and elevations, now filed in the Archive of the Indies in Seville, was ambitious. It called for the construction of a teaching hall, an herbarium, a natural history museum (*gavinete*), living quarters for Cervantes and the staff, and a research library open to the public, plus an enclosed teaching garden with colonnades. Beyond these buildings lay the massive main garden, which was to be encircled by an irrigation ditch and a stone wall and accessed by a bridge leading into the main gate.

Despite its ambitious size and complex infrastructural demands, the site plan by Constanzó is remarkable for its intentional visual simplicity, itself a gesture to the rational, scientific ideal it represented (see map 3). The grounds were to be divided into approximately thirty beds, each almost as large as a city block. The recently refurbished Chapultepec aqueduct on the south end of the garden was to provide water, channeled through the canal around the perimeter and then into a grid of pipes that extended along paths throughout the garden. Water from the aqueduct was to be supplemented in the dry season by a pond in the southwestern corner of the garden, a large fountain near the center, and twenty-four smaller fountains for individual beds.[38]

The planned garden shared characteristics with European royal gardens of the era more than it might seem at first glance. Typically, an eighteenth-century royal garden mapped its patron's territorial domination. In Versailles, for instance, land nearest the palace emphasized legibility, rationality, and the centrality of regal authority. As visitors ventured farther afield, they moved into less rigidly controlled spaces that symbolized the broad territorial reach of the French king's power. Kew Botanical Garden outside London, meantime, used its layout, architectural structures, and plants to map England's dominion over tropical resources encircling the globe. Madrid's Royal Botanical Garden similarly asserted the Spanish king's claim over the products found in his territories around the world.[39] The Mexican garden, austerely laid out in uniform blocks, seemed to lack these standard tropes, which might give the false impression that it ignored the organization of power, space, or knowledge. Yet the garden did comment on such matters through its promise to impose top-down order and rationality on to a colonial context that resisted such impositions.

This was an expensive proposal. The Crown had agreed to cover all the expedition's salary and travel expenses but expected the garden and the school, like other royal institutions, to become self-sustaining. With no clear source of income other than royal funding, which the Crown was loath to guarantee, Sessé and Cervantes had to scale back their plans. As head of the expedition, it fell to Sessé to figure out not just how to pay for the construction costs, but also how to make the garden's scientific research profitable. This task was made all the more difficult owing to the fact that Sessé now was spending most of his time conducting field research in rural sections of the colony. Barring the discovery and rapid development of a wonder drug equal to Peru's cinchona (used to treat malaria), Sessé had to seek sources outside science. Institutions typically benefitted from rents, fees, and taxes, but most of these, including even the licensing fees paid by pharmacists, were already claimed by other agencies. Among these other claimants was the Mexican medical licensing board, which Sessé already was at odds with and did not wish to antagonize further. He explored a plan to build a bullfighting ring to fund the garden, and even commissioned architectural plans and elevation drawings, but the startup costs were prohibitive, and it turned out that the income would have been inadequate. He secured patronage from members of the New Spanish elite, but this was not enough to cover more than a fraction of the expenses.[40] After much pleading, Sessé persuaded the Crown to support the garden while he continued his search for alternative sources of funding. No one predicted at that time that the garden and school would remain permanent dependencies of the state treasury.

Financial and bureaucratic difficulties faded in the face of a more pressing dilemma. It turned out that seasonal flooding and poor drainage not only made it difficult for students to attend class or work on their experiments but also rendered the entire Potrero de Atlampa unsuitable for gardening. The now-famous Trasmonte map, created over a century earlier in 1628, depicted what Mexico City could look like if the government were to make progress on its desagüe project.[41] Even in this optimistic view, the Potrero de Atlampa in the map's foreground appears uninhabitable (see the footnote for a link to a high-resolution dynamic version of this map).[42] Though the on-going desagüe project had improved the situation during the century and a half since the time of the Trasmonte map, the land in the Potrero de Atlampa remained marshy. It was with good reason that the common name for the Potrero de Atlampa was "El Sapo," an allusion to the toads and frogs that enjoyed its soggy conditions. But Sessé confidently insisted that it couldn't be any worse than the soil of gardens he had known in Spain. Moreover, he had been seduced by the site's proximity to the city's center, and by the ditches that bordered parts of the lot, which he hoped could be used to drain the garden, and which also seemed ideal for aquatic plants.[43]

As the other arms of the expedition forged ahead with fieldwork, clinical trials of simples, training the next generation of Mexican doctors and pharmacists, and assembling an herbarium, the garden remained, literally, mired in the mud. Cervantes fretted when flooding suffocated plant roots and rotted seeds that had been transplanted from the far corners of the New Spanish countryside or acquired at great expense from Europe and Nueva Granada (modern-day Colombia, Ecuador, Panama, and Venezuela). This was compounded by the lack of the sandy loam that every gardener cherishes. Instead, the ground was heavy clay that remained so saturated after the flood water receded that students and gardeners could not walk on it without their feet getting sucked under, much less till or plant it. When the ground finally dried, it turned hard as stone until it flooded again. One can only imagine the misery of the garden assistants and peons housed in *jacales* (wattle-and-daub shacks) hidden away in the least hospitable corners of the property.[44] The garden, armed with the latest tools of Enlightenment science, was supposed to affirm man's domination over the natural world, forcing it to conform to abstract schemes of rationality. But nature was presenting unanticipated resistance.

As a stop-gap measure, Cervantes set up a provisional four-acre garden surrounded by a picket fence in the patio of the temporary quarters the school rented from Ignacio Castera, where he had the garden workers elevate almost two thousand flowerpots atop wooden racks. These pots worked reasonably well for small plants but could not accommodate the trees that Sessé and other members of the expedition had begun sending to the garden from the countryside. In the meantime, Cervantes hired plowmen to prepare two of the Atlampa garden's large sections and pleaded with the Crown to send to Mexico a head gardener trained in the latest methods by Gómez Ortega at the Real Jardín Botánico de Madrid—it would not be until two years later, in 1790, that the Crown finally appointed Jacinto López to the role.[45]

Despite these setbacks, Cervantes remained resolute that modern science must triumph over nature, even if this meant begrudgingly borrowing aspects of native technology. To maintain the fertility of their gardens (known as chinampas), which were surrounded by water, indigenous farmers south of the city regularly dredged soil from the lake bottom, rich in decaying matter, and piled it atop the chinampas for fertilization. These indigenous farmers stabilized their plots by planting deep-rooting ahuehuete trees, and maintained a solid section in the center on which to build a home.[46] Impressed by the ability of chinampas—which the Creole intellectual Clavijero had confusingly coined "floating gardens"—to prosper in flooded conditions, Cervantes in 1788 and again in 1789 commissioned raised beds and bought hundreds of canoe loads of high-quality soil dredged from the lake bottom in Tlalpan.

The problem was that this strategy ignored aspects of the environmental conditions of Atlampa. Chinampas were long narrow mounds built on many layers of buoyant reed woven into braids, anchored in the shallow lake by wooden stakes. These features allowed effective drainage and continuous filtration, preventing waterlogging or suffocation of the plant roots. The Royal Botanical Garden in the Potrero de Atlampa borrowed only one part of the native methods: the use of dredged lake soil. Mounding this nutrient-rich soil atop the existing heavy clay did nothing to address the problem of poor drainage.

Try as they might, Cervantes and the site planner, Miguel Constanzó, could not bend the land to their will. Not even the costly installation of wooden drainage pipes, nor attempts to deeply plow the land made any difference.[47] The expedition had expected to use the garden to propagate plants plucked from all across New Spain, whose ecological zones ranged from temperate woodlands to desert, from coastal to alpine, and from rainforest to chaparral. However, nothing but the hardiest of weeds could survive these conditions. They could not even get maize to grow. Cervantes and Sessé at last accepted failure at El Sapo.

During their search for a better location, they eyed the site of Moctezuma's preconquest garden on Chapultepec hill, west of the city. After the conquest, the Spanish had abandoned the Aztec garden. In 1784, the viceroyalty of New Spain had begun construction of a sprawling palace on the site. The building, today known as Chapultepec Castle, sat incomplete at the time of the Royal Botanical Expedition, with the government unable to finish it or find a private buyer. Were it not for the archival work by the Creole scholar Francisco Javier Clavijero, who studied the texts of the conquerors to reconstruct knowledge about Aztec gardens for an Enlightenment-era audience, the older history of the location might have been forgotten.[48] Sessé argued that the grounds surrounding the abandoned facility would be ideal for the Royal Botanical Garden. The structure could be finished to fill the needs of the garden and the school, and the many levels of elevation and microclimates of Chapultepec would permit the cultivation of a wide variety of plants gathered from across New Spain.[49] Once again, members of the Royal Botanical Expedition were turning to indigenous and creole knowledge that they claimed had no place in modern science. Their proposal to move the garden to Chapultepec was a great idea, but it was rejected by the government.

IMPERIAL SCIENCE OVER MEXICAN NATURE

Despite its initial tensions with the creole establishment, the expedition soon earned an enduring place within Mexican society. Its reforms to the study and licensing of pharmacy, medicine, and surgery were gaining acceptance by the early 1790s. Its alumni started entering the professional world. And the expedition's quick success at identifying hundreds of medicinally useful plants was attracting attention. But the garden where these plants were to be propagated and where students were supposed to conduct their research was still stuck in the mud of Atlampa.[50]

In 1791, Cervantes and Sessé finally won an alternative space for the Royal Botanical Garden, not near the unfinished castle atop Chapultepec but within the highly prized viceregal palace on the Plaza Mayor, an unrivaled viceregal endorsement of the endeavor. Installation of the garden required remodeling the southeastern quadrant of the royal residence, including conversion of some of the existing arcades into a greenhouse and others into teaching laboratories for medicine, surgery, and pharmacy. This placed the garden right next to the Royal Mint, which occupied the northeastern corner of the palace. A move to the viceregal palace dramatically reduced the garden's construction and maintenance costs, while conveniently shifting the remaining expenses directly to the viceroy. The patio could hold only one thousand species, but this was sufficient for the school's needs, provided that these were supplemented with plants collected daily from the countryside or shuttled back and forth between the viceregal palace and Chapultepec (a section of which Sessé and Constanzó finally gained permission to use). The downtown location, moreover, was near housing for staff and offered easy access to teaching hospitals for Cervantes, Sessé, Mociño, and the students. It also was a publicly visible space from which to edify colonial subjects regarding the benefits of Spain's Enlightened absolutism.

By bringing the garden into the center of colonial authority in New Spain, the expedition, the viceroy, and the Crown expected that the union of absolutist politics and Enlightenment science could overcome the obstacles set forth by nature so that they at last could get on with the business of reorganizing space, knowledge, and power. As it turns out, the structure could be easily remodeled to meet the school's teaching and research needs, but the land, once again, presented a daunting challenge.[51]

The soil was an accumulation of clay mixed with rubble from the old Aztec temples that had been toppled by the conquistadores. Scarred by his battle against Atlampa, Cervantes insisted to the viceroy that altering the height of the beds or covering them with good soil would not be enough: the entire site had to be excavated to a depth of at least one *vara* (around one meter) so that the clay and rubble could be discarded and replaced with fertile, well-draining soil from Tlalpan, along with underground drainage pipes.[52] These subterranean wooden pipes, which seem to have drained into the Acequia Real that ran along the garden's exterior south wall, imitated the kind of filtration that layers of woven reed nets created for chinampas. By harnessing Enlightenment science and imperial power, Cervantes literally and metaphorically sought to replace the historical accumulation with new fertile soil amenable to the appropriation of Mexican nature.

This massive earthmoving was accomplished by hundreds of indigenous laborers working under the supervision of the Spanish head gardener Jacinto López. An endless procession of native porters in canoes floated past the palace through the Acequia Real—a section of the old Aztec canal system that still served as part of Mexico City's transportation system. The canoes brought in fertile soil of the sort used for chinampas, along with tools, bricks, tiles, lime, wood, and other building material, and took away the old soil and preconquest rubble.[53] While there are no images depicting the undertaking, a lithograph by the Mexican artist Casimiro Castro shown in figure 11 gives an idea of how the canals connected rural people and resources to the urban center. As shown in Castro's lithograph from the 1850s, canals continued to serve as vital transportation arteries in central Mexico City even six decades after the viceregal garden project.

Cervantes praised Jacinto López for conceiving of and executing the project.[54] By doing so, he avoided acknowledging his dependence on indigenous knowledge. The garden relied on a small army of native gardeners, including such full-time indigenous cultivators as Mariano Ylario, Diego Martín de la Cruz, Juan de la Cruz, José Ramón, Mariano de la Cruz, José de los Santos, and Antonio Trinidad. López, by contrast, was a Spaniard who had trained under Gómez Ortega and worked at the Royal Botanical

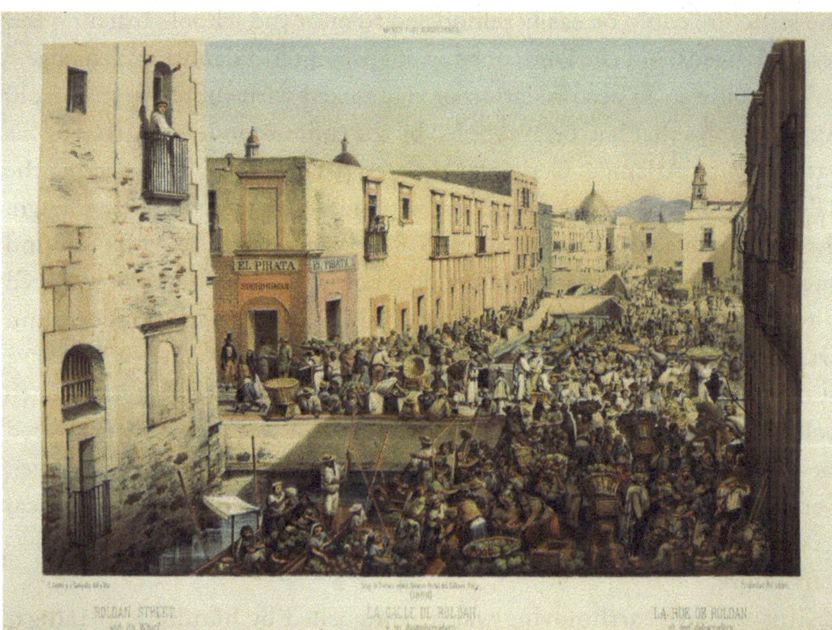

FIGURE 11 Casimiro Castro, *La Calle de Roldán y su desembarcadero*, in *México y sus alrededores, 2nd edition* (Mexico· Imprenta Litográfico de Decaen, 1864 [1855]).

Garden in Madrid for fourteen years before being hired by the expedition in 1790.[55] By praising López with his European training, rather than indigenous people with their local experience, Cervantes maintained his commitment to Enlightenment science while obfuscating the degree to which he had come to rely on local knowledge.

The two parts of the garden, one in the downtown viceregal palace and the other atop Chapultepec, worked in tandem. As Sessé explained, the beds in Chapultepec grew "the plants from across the diverse climes of this America," while the viceregal site offered students and teachers a space for research and pedagogy, and to the public it offered "the Capital its most beautiful spectacle, visible from its principal square."[56]

Sessé insisted that *gente decente* (respectable people) be allowed to enter the garden within the viceregal palace, and he asked that the southern wall separating the garden from the street and canal be shortened so that the public could admire the plant collection from beyond the barrier. When government officials questioned whether this was a proper function of the garden, Sessé pointed to Article 4 of the Reglamento del Jardín, which called on it to serve a public function.[57] By allowing subjects from all walks of life to witness the garden's rational regimentation and its control over Mexican plants, Sessé wanted to demonstrate the ability of Enlightenment science, backed by the power of the state, to control nature within the colony. Inside the walls of the garden, the school of botany and medicine taught not just any science, but a form of science that directly asserted the authority of Mexico City over all New Spain, and that linked the colony to the empire.

The political need to engage the public sometimes came into conflict with the scientific goals of the garden. This tension bubbled up even with seemingly trivial matters, such as the question of where to place garden paths. To maximize the use of the expensively installed soil, Cervantes resisted the viceroy's plan for an elaborate network of strolling paths for wealthy guests. Instead, he asked that paths be placed only where they were most essential. His request to limit the paths to maximize the use of the soil paid for by the royal treasury became a convenient way for Cervantes to affirm the scientific priorities and his own role as gatekeeper, presented as an effort to maximize the viceroy's own financial investment.

Tension between the scientific and political goals of the garden also had recently blown into open conflict during a public lecture, when members of the expedition insisted on receiving the viceroy's regent as a compatriot of science rather than with the typical pomp and ritual. The regent departed the garden in a huff, and the school of botany and medicine was forced to publicly apologize and to restage the entire event according to protocol. Cervantes, therefore, had to tread lightly in this latest effort to prevent the Royal Botanical Garden from following the model of European botanical gardens, whose scientific purposes were often overshadowed by the demand that they serve as pleasure gardens for the economic and political elite.[58]

CONCLUSION

In their pursuit of control over Mexican nature, Sessé and Cervantes patted themselves on the back while obfuscating the fact that they had been forced to accept limitations imposed by nature and had borrowed indigenous knowledge. The Crown, however, would have none of it. In late 1792, after Sessé and Cervantes had settled into the viceregal palace and set up auxiliary beds in Chapultepec, the king's minister fired off an irate missive excoriating the members of the expedition for failing to prevail over the obstacles presented by the Potrero de Atlampa, impugning the abilities and integrity of Sessé, Cervantes, and even Constanzó. He ordered them to move the garden back to the Potrero de Atlampa and bend nature to their will.

Sessé calmly responded with an extended review of the Chapultepec site, examining its woodlands, ditches, and slopes and its ready access to water. He explained how the variety of settings it offered enabled the expedition to naturalize plants from such diverse places as deserts, forests, marshes, and even alpine regions. All this would be impossible in Atlampa but could be accomplished in Chapultepec, he explained, while avoiding the high cost of irrigation and drainage. To make his case as forcefully as possible he reminded the minister that only by cultivating this wide variety of species could the garden find ways to propagate and exploit the medical and commercial potential of New Spanish plants for the benefit of the empire.[59] Even the viceroy wrote the Crown in support of Sessé, arguing that, while Chapultepec might lack the attributes needed for an agricultural farm, it offers the variety of "flats, hills, and slopes" that made it ideal for the botanical garden.[60] In the end, Sessé prevailed through a deft combination of argument, obstinacy, and mobilization of alliances in Madrid and Mexico City.

Firmly established in the viceregal palace and Chapultepec, the Royal Botanical Garden bestowed two commonalities on the plants that moved through its beds. First, almost all of them possessed potential economic and scientific value that could serve both the colony centered in Mexico City and the empire based in Madrid. Second, except for the handful of plants explicitly imported from Europe, Asia, or South America, they all came from unspecified locations within the territory claimed by New Spain. Members of the expedition did not concern themselves with plants' potential value to local people. Neither did they record the location where they collected particular specimens. All that mattered was that the plants had potential economic and scientific value to the empire, that they had been collected within New Spain, and that they now were in the possession of the expedition.

The expedition's earthmoving projects and its centralization of plants from across the colony might not have been as visually striking or awe-inspiring as the terraces of Versailles or the palms pressing against the ceilings of massive glass greenhouses in Kew and Madrid, but they were no less an achievement. Among the accomplishments was the erasure of the indigenous foundations on which the garden was built. It quite literally excavated and discarded the historical accumulation from the site of its construction, starting afresh with the products of Enlightenment science and imperial power—behind which indigenous knowledge was hidden—to create a fertile soil amenable to the appropriation and reorganization of Mexican nature. This power to render invisible the indigenous knowledge on which it depended, in fact, was central to the garden's Enlightenment mission.

In the end, Cervantes and Sessé traded visual awe for a studied admiration of the ability of men of science, backed by the Crown and New Spanish supporters, to make the land look pretty much as it had before. That, after all, was what they hoped to do with New Spain: keep it looking as it did, growing what it grew, and producing what it produced, but to do so better, under scientific supervision, and for the benefit of the empire. And they hoped to concentrate products from the vast, diverse spaces of New Spain into one centralized hub, whose organization appeared as though it were subject only to the abstract rule of Enlightenment rationality.

CHAPTER 4

ILLUSTRATION ON THE MEXICAN BOTANICAL FRONTIER

When King Charles III ordered the Royal Botanical Expedition in 1787 to take the findings from Francisco Hernández's Renaissance-era Mexican expedition and bring them up to date "with the current state" of Enlightenment science, he was referring mostly to the scientific system developed by and around the Swedish botanist Carl Linnaeus (1707–1778).[1] This chapter examines what that Enlightenment-era system was, how members of the Royal Botanical Expedition put it into practice, and the central role visual production played in this process. Most of all, it addresses how members of the expedition used Linnaeus's system—as well as how and why they sometimes seemed to flagrantly violate it—to move beyond the work of Hernández, as they searched for plants rooted in colonized soil that could be exploited for the benefit of the Spanish Empire. It also shows the implications of their approach for their construction of a particular understanding of New Spanish nature, and the unanticipated effects their methods had on Mexican society.

IDENTIFYING A FLOWER, AND WHY IT MATTERS

We live in an era when the ideas advocated by Carl Linnaeus have become so normalized it can be difficult to recall what was radical and powerful about his vision. To illustrate its importance, we might start by considering a yellow flower in a meadow. If we bend down to examine the flower sticking its head up through the grass, how do we know what we are looking at? Is it a dandelion, or another flower that looks like a dandelion? Could it be a hawkweed that, through some series of accidents—a deformation caused by a fungus, for example, or perhaps mutilation by an insect or a lawnmower—came to resemble a dandelion? If it is a dandelion, what kind of dandelion is it? And is the similar-looking flower growing near it a dandelion of the same kind? If we are collecting dandelions for pharmaceutical reasons and, supposing that one kind might offer a certain medical benefit but another kind does not, how do we make certain we are collecting the right kind? And, if we are scientists exploring uncharted botanical territory, how can we determine whether it is the well-known "common dandelion," or some new species? In short, how do we know what kind of plant it is? These are questions that vexed scientists (that is, naturalists) for hundreds, even thousands, of years.

During the Enlightenment, European naturalists, eager to bring intellectual order to the products of nature, sought to identify clear boundaries between one kind of plant and another, boundaries that would hold true no matter where in the globe a plant might be found. This led to the sorting of living things into stable categories and subcategories—running from domain at the top all the way down through kingdom, phylum, class, order, family, genus, species, and variety. These categories are known as taxa. We tend to feel comfortable assigning specimens to big categories, such as the animal kingdom or plant kingdom, but our confidence may fade when we get into finer distinctions. Thus, when we speak of plant taxa, we usually are concerned with the lower levels: genus, species, and variety, where there is greatest uncertainty and possibility for error.

During the Enlightenment, naturalists, following the lead of Linnaeus, initiated the practice of designating particular specimen samples as those against which all others should be compared to confirm their species and variety. Each such specimen sample was known as a "type specimen." And each type specimen was assigned a Latinate scientific name, or binomial.[2] Thanks to this innovation, if I want to determine whether the dandelion that I found in the meadow is really a "common dandelion" (*Taraxacum officinale*) rather than, say, a "red-seeded dandelion" (*Taraxacum erythrospermum*), I would compare its reproductive organs (such as the stamen and pistil) to those of the designated "type specimen" for the *Taraxacum officinale* of the Asteraceae family. The official type specimen is kept in an herbarium—a plant-specimen library—somewhere in the world, such as the Muséum National d'Histoire Naturelle in Paris, the New York Botanical Garden in New York City, or the Real Jardín Botánico in Madrid.

To make things slightly more complicated, in place of a complete sample, the original type specimen of a plant might just be a fragment or a drawing, or even a drawing of a plant fragment. This is true, for example, for many of the plants from the Hernández expedition and the Royal Botanical Expedition, which were undertaken during an era when collection, transportation, and preservation of samples was difficult. Moreover, since botanical research during the Enlightenment occurred as a mad frenzy rather than as a slow, controlled process, it is common to find that many people collected and described the same plant at around the same time, with irresolvable disputes over who was the true discoverer. It also might occur when an expedition drew two different specimens that, in retrospect, appear to represent the same species. In cases where such resolution is impractical, several of the oldest samples or drawings are designated as "syntype specimens."[3]

Linnaeus's system has been refined over time but still guides common practice—though aspects of it have been challenged in recent decades by phylogenetic nomenclature based on genetic analysis of ancestry. As biologist Jay Withgott has pointed out: "In the history of science, few innovations have had as wide or as lasting an impact as the system Swedish naturalist Carl von Linné (Carolus Linnaeus) developed to name and classify the world's organisms. Imposing order on a lawless proliferation of names, Linnaeus created a system 250 years ago that, with substantial modification, still stands today."[4]

While all this business about Linnaean classification can seem arcane, there are three key points I want to emphasize. First, to create a sense of certainty about how they organized items within the natural world, Enlightenment scientists developed an ingenious method of organizing plants into a universal system based on each flower's reproductive organs. This system guided botanists and researchers whether they were digging up plants in local environs, exploring the far corners of the world, or assigning evolving sets of attributes to a plant. Second, we should not confuse this man-made system for the actual structure inherent to nature. Its emergence grew out of a historical process, and this process will be part of the subject of this chapter.[5] Finally, at every step of the development of this system, scientists relied on particular practices of visualization. This chapter examines these practices, and how they helped imperial agents in New Spain to understand, lay claim to, and exploit the natural world around them.

HOW IT CAME ABOUT: RESOLVING "THE CHAOS OF NOMENCLATURE" THROUGH VISUALIZATION

In the 1570s, Hernández had tried to solve "the chaos of nomenclature" that plagued pre-Enlightenment botany by introducing a universal taxonomic system based on Náhuatl.[6] But his system was poorly conveyed by the editors of his manuscripts, and the little that other scientists did learn about his original ideas struck them as weird and confusing. In 1583, the Florentine natural historian Andrea Cesalpino (1519–1603) published *De plantis libri XVI*, in which he proposed his own system of plant taxonomy. Impressed by the ways armies were subdivided into ever smaller units—regiment, battalion, company, platoon, team, and so forth down to the individual soldier—Cesalpino proposed a similar structure for organizing plants. At the time, Cesalpino's system was just one of many that were floating around. Its real importance only came later, when Carl Linnaeus drew inspiration from its organizational structure and use of flower's reproductive organs as a basis for differentiation.[7]

Before Linnaeus, botany concerned itself with disorganized and competing sets of concerns about the general nature of the plant, how it grew, its medicinal uses, how it emerged from the soil, and so forth. When Linnaeus came on the scene starting in 1735 with the publication of his *Systema Naturae*, his great innovation was not the introduction of the Latinate binomial, which had been used haphazardly for several centuries, but rather, as stated by Anna Pavord, his insistence that "all a name had to do was designate. It did not have to describe."[8] He also imposed standardization, achieved by establishing himself as the hub of the system, assisted by legions of loyal disciples who promulgated his ideas with an unrivaled evangelical intensity and global reach. Of equal importance, as pointed out by Londa Schiebinger, Linnaeus "taught that the purpose of natural history was to render service to the state," whose "wealth could be aggrandized through the exact study of nature."[9] Linnaeus and his followers transformed botany into a premier science of the Enlightenment and of European imperialism by imposing a system that seemed, at last, to bring rational standardization to the overwhelming amount of information that had become available, and centering that knowledge within European states and imperial institutions. Rather than a mere alternative or supplement to existing practices, the system promoted by Linnaeus and his disciples was supposed to supplant all that had come before.

Without completely dismissing the preoccupations of his predecessors, Linnaeus subordinated their concerns to his classificatory scheme based on visual observation. He created a universal system into which one could fit any known or newly discovered plant by zeroing in on the appearance of the organs that seemed most essential to reproduction. As they added new plants to the system, Linnaeus and his adherents assigned each one a Latinate binomial, typically honoring a male European botanist, that superseded all previous scientific and common names. As Beth Fowkes Tobin notes, by ignoring "plant anatomy, physiology, and ecology," Linnaeus created a manageable, but very narrow, view of nature.[10] The success of this system relied on the coordinated efforts among a tight network of scientists based in European centers who exchanged plants and knowledge with agents and colleagues around the world. These exchanges occurred through the literal transfer of live or dried plants and by sharing illustrations rendered by specially trained draftsmen.

Visuality was central to the Linnaean system, which was based on observable details, especially regarding a plant's sexual organs.[11] Herbaria and gardens gained heightened importance as tools of visualization within his method. But it was botanical illustration that played the starring role.

Until Linnaeus, uncertainty about how plants should be organized or named contributed to a lack of consistency in how they were illustrated. He solved this by instructing artists to eliminate everything except for the information required by his system of classification. By removing plants from their context, Linnaeus and his acolytes defined them as completely transferable and open to being possessed by imperial centers without counterclaimants. Botanical illustrations empowered the followers of Linnaeus to reliably monopolize the power to determine how individual plants were viewed, categorized, collected, claimed, and disseminated globally. The conventions of modern botanical illustration that we take for granted today have their roots in these innovations introduced by Linnaeus and standardized by his many disciples.

THE ILLUSION OF MIMESIS

Enlightenment science placed enormous importance on visual images, but only if those images followed rigid guidelines. An Enlightenment-era botanical drawing, in fact, might be better understood as an illustration of a concept than of a plant. As draftsmen adapted their eyes to see through what I term a "Linnaean optic," they learned to manipulate their images and erase everything that was irrelevant to that system. The resulting illustrations followed specific rules. Against a white background, they depicted a plant's flowers, leaves, and perhaps a snippet of a stem to which the flowers or leaves were attached. Near the top or bottom border, they added details of fruits and flowers, shown at different stages of growth or dissection. Just as important as what these images included was what they excluded. They offered no indication of the plant's size or growth habit and usually excluded the roots. While these images depicted dissected flowers and fruits, they did not show dissections of any other part of the plant, such as stems, leaves, or roots. They also excluded any information about the plant's ecosystem, soil, location, climate, or season of growth, or its relationship to people or animals.[12] As we will see later in the chapter, draftsmen from the Royal Botanical Expedition to New Spain usually complied with these norms, but at other times they brazenly violated them, for reasons and with implications that will be discussed.

Botanical illustration became a more powerful tool than either a botanical garden or dried specimens in an herbarium. A live plant in a greenhouse or a dried one pressed between sheets of paper enabled a scientist to observe one specimen at a specific moment in its life cycle, but only botanical images and written texts could collapse different stages of growth and multiple specimens—observed and depicted over and over again across many years. And only botanical illustrations and texts could distill the information to the type's essential elements, erasing all distractions and aberrations.

While Linnaean botanical illustrations give the illusion of heightened accuracy and a truer mimesis of nature than still-life paintings, in fact they required greater distortion and manipulation of the relationship between the artist's model and his drawing. To Linnaean botanists these departures from mimicry were seen as truer, more accurate depictions of nature because, rather than be distracted by what was in front of the artist's eyes, they portrayed the underlying structures of knowledge that had been unlocked by Enlightenment science. To emphasize things that would not stand out to one not trained in the Linnaean system, the artist might exaggerate the size or color of a flower or its reproductive organs to make their details more visible than they might be in real life. They might depict a plant with multiple flowers, each at a different stage of florescence to show specific features that botanists had defined as characteristic of that species, even if in reality all the flowers on the specimen would have been in the same stage of blooming. Or else the draftsman would be required to show the plant with a particular stage of leaf growth and a particular stage of flower growth as though they occurred simultaneously, even if in real life they would have occurred at different times of the year.[13]

Isolation of the specimen, or parts of the specimen, against a white background signaled to the viewer that this was a scientific illustration rather than an artistic painting. The practice of isolating specimens on a white background was not new. Since the sixteenth century it had become established as a central component of a nature study. As historian Janice Neri has pointed out, "specimen logic," which became popular during the sixteenth century, "turns nature into object by decontextualizing select creatures and items—that is, by removing them from their habitats, environment, and settings. Conversely, only those creatures or items that can be depicted or displayed as objects, those that possess clearly defined edges or contours and whose surfaces are visually distinct, are suited to the aims of specimen logic."[14] While Enlightenment-era botanical illustrators did not invent "specimen logic," they gave it greater importance as they isolated details to emphasize features that Linnaean botanists decided were characteristic of the particular species.

When artists who had been trained in still-life painting transferred their skills to botanical illustration, they continued to rely on the practice of close observation that they had learned in the art academy, but in place of the particular, the ephemeral, and the palpable, they dwelled on the ideal type, the timeless frozen moment, and the plant as idea rather than tangible reality. Unlike a still-life painting, which typically highlights each flower's individuality and gradual decay, and plays with light and shadow, Linnaean illustrations show the various stages of life and growth, not as the passage of time, but as collapsed moments that become unified and eternal, with no hint of the depressing decay that follows fructification. And it is all well lit, from every direction, without deep shadows that might obscure important details. The purportedly accurate botanical illustration, therefore, is a representation of something that would not be found in nature. This divorce from reality was not a problem for Linnaeus because, as Beth Fowkes Tobin points out, his "system of classification pivoted on the concept of the ideal type, and the true vegetable kingdom existed, for him, in an ideal place divorced from the particular, the local, and the agency of time."[15]

To accomplish these tasks, an artist had to be rigorously trained in art and in the Linnaean system, and strictly supervised by a botanist. The Linnaean system, therefore, prioritized visuality, but, ironically, this visuality relied on a fictional interpretation of what one might behold. It conditioned illustrators and scientists, and even aficionados, not to trust what they saw with their naked eye, but instead to "see" in a particular way that was far from natural.

TRAINING LINNAEAN DRAFTSMEN AND ADAPTING THAT TRAINING TO THE MEXICAN BOTANICAL FRONTIER

Artists who succeeded in cultivating a "Linnaean optic" found success, and even fame, in their new profession. But how did they cultivate these skills? And how did they adapt them to the challenges of exploring Mexico's botanical frontier?

Typically, training as a botanical draftsman in Europe unfolded as a two-step career. First, one completed a traditional academic training in painting, most often in an academy, but other times with a mentor in a painting workshop. Second, artists who excelled might be recruited by a botanist who tutored them in Linnaean theory while requiring them to copy and reconcile existing botanical illustrations, comparing multiple versions to refine their ability to create their own version that conformed to the expected characteristics of particular taxa. Once the artists had learned to copy and reconcile existing illustrations, they then learned to compare them to a specimen at hand, to look for any small details that could help refine our understanding of the essential characteristics of that specimen's taxon.

During this process, the botanist hovered over the draftsman's shoulder and gave specific feedback about ways to more effectively bring the artist's eye and the resulting image into line with Linnaean expectations. As one eighteenth-century botanist explained about his relationship to his draftsman: "I told him which parts of a flower and fruit should be presented particularly clearly so that the character of its genus could be clearly recognized."[16] Another European botanist described his need to teach his draftsman the theory and even dissect the plant with him so "that the different parts, whether in their natural size or enlarged, could be better depicted by him."[17]

In addition to their work with botanists, European draftsmen of the eighteenth century consulted training manuals. These explained the Linnaean system, offered practical advice, and even cautioned artists about which details of a specimen they should discount. One manual, for example, reminded artists to ignore double flowers, since these were accidental disfigurements, and the truth of "nature will no longer be found among them."[18] Though double flowers might be products of nature and among the most visible features of the plant, as well as an inheritable trait quite common among certain species (e.g., zinnia, rose, gardenia, hibiscus, carnation, and peony), they were not to be considered part of "nature," as understood through the Linnaean optic, and therefore should never be included in a botanical illustration.

One of the most popular manuals was James Sowerby's *Botanical Drawing-Book,* published in 1788, based on his personal experience training draftsmen. Assuming that the reader had already mastered painting, the manual directly entered into an explanation of the Linnaean system and the seven fructification organs of a flower. Then it showed the expected variations in the shape of each of these seven organs so that the artist could know what to look for and then depict what he saw in conformity with Linnaean expectations. Like all such manuals, Sowerby's charts of expected variations were based on those originally published by Linnaeus in his landmark *Systema Naturae* in 1735—which had been drawn by Linnaeus's collaborator George Dionysius Ehret (1708–1770). If the artist did not see his specimen as conforming to one of these examples, then he was not looking at it correctly and must adjust his vision. Sowerby concluded by offering advice about useful pigments and how best to apply them to exaggerate the features deemed important by the Linnaean system. Thus armed, the draftsman could recognize acceptable variations and convey this information unambiguously through the painted image.[19]

Even experienced draftsmen maintained the practice of constantly copying and recopying existing illustrations to discipline their eye and hand to conform with expectations. Every time they drew a plant, their image became yet another visual testament to the Linnaean claim that every plant occupied a predetermined place within his universal system. The more widely these kinds of images were disseminated, and the greater the number of artists who learned to produce them, the more these botanical illustrations became accepted as accurate depictions of nature, and the more the Linnaean system became ingrained as common sense.

Artists were guided not just by existing images and by their trained eye, but also by texts. While much has been made of Linnaeus's binomial naming system, Linnaeus considered these names to be merely handy abbreviations for everyday use. What he actually found most useful were the long descriptive names he gave to the plants. For example, his long name for sweet vernal grass, *Anthoxanthum odoratum*, was "vernal grass with an ovoid-oblong spike and florets on short peduncles that are longer than the awn."[20] Biologist Maura Flannery explains that what Linnaeus accomplished with this practice of assigning both a short binomial and a longer descriptive name "was to separate the name from the species description, so the name did not have to do the work of both."[21]

For known and well-described species such as sweet vernal grass, the illustrator strove to show the details described by the long name in exaggerated clarity, aided by previous illustrations of the species. But the situation became all together more complicated for members of the Royal Botanical Expedition to New Spain, since part of their mission was to draw previously unknown plants, which had not yet been assigned a long name or a short name, and for which there was no collection of previous descriptions or illustrations to guide them.

Members of the expedition encountered thousands of potentially new taxa during their almost seventeen-year expedition and had to reconcile these potentially novel plants with the work of Hernández and of other European botanists. How, then, did they ascertain in the field whether a specimen was new? And how did they figure out what they should draw when they encountered a plant that they suspected had never been described or illustrated before, and which lacked a Latinate binomial abbreviation, much less a long-form Linnaean description or a paper trail of previous depictions? What guidance did they follow to discern which of the features they observed in their specimen were typical of the species, and thus should be emphasized, and which were aberrational, and therefore should be discounted? How did they decide what additional details to include in their drawings to aid botanists who would later be tasked with slotting the new plant into the existing Linnaean system? The rest of this chapter addresses the ways the members of the Royal Botanical Expedition to New Spain navigated these challenges.

ILLUSTRATORS FOR THE ROYAL BOTANICAL EXPEDITION

Vicente Cervantes explained to his students in Mexico City in 1793 that "knowing the language of botany, which comes about only through a clear understanding of the composition and connections among the classes and orders of the [Linnaean] system and learning to name and describe, through art, all the parts of each plant, one can easily read the great book of nature."[22] Cervantes and his students, along with other members of the Royal Botanical Expedition, saw themselves as the vanguard in the search for new knowledge in the service of empire. To succeed at their task, it was imperative that they learn to wield Linnaean botany within the fields and forests, as well as the pharmacies and hospitals, across New Spain. For this knowledge was to be their most powerful weapon for revealing and harnessing the mysteries of nature.

As Cervantes made clear, visual observation, description, and depiction were at the center of this endeavor. On the surface, the practice of Linnaean botany in Mexico mirrored that of Europe, but reproduction of metropolitan practices in the colony was not a straightforward process, especially since the purpose of this reproduction was to describe things that had never before been scientifically described.

The Crown's orders chartering the Royal Botanical Expedition included four lengthy supplements, likely drafted by Casimiro Gómez Ortega, director of the Royal Botanical Garden in Madrid, that offered a remarkable level of detail regarding exactly how each arm of the expedition should undertake its tasks. Taken together, these royal orders demonstrate the tight connection among the branches of the expedition, as well as their dependence on visual practices and their relationship to local indigenous knowledge.

The first letter was directed to the botanists and naturalists of the expedition. It ordered them to keep an eye out for plants that might be new to science and to send all their work and any plants they collected to the garden. The instructions did not tell them to record the location where they collect their specimens, but it did require that they keep a written log of the plants they collected, including any names the plants might carry in Spanish or in the various indigenous languages. And they must write down any qualities that informants, "even . . . Indians and Locals [*Indios y Paysanos*]" attributed to the species.[23]

While the members did the collecting, it was the garden's school of botany and medicine that would be ultimately responsible for proposing how to slot each plant into the Linnaean system. The letter of instructions to the school started by providing a detailed recapitulation of the Linnaean system it must follow. It also stipulated the textbook the school must use, noting that it was the exact same one used in the school at the Royal Botanical Garden in Madrid.[24] It then described how the school should assign names, in light of the fact that the expedition was in search of previously unknown plants: "In the discovery of New Plants . . . assigning names is particularly important to fix the understanding of which plant one is talking about so that its purported characteristics can be examined, with false ones rejected and new virtues properly assigned. This also will aid in attempts to acclimatize the plant to places where it does not grow naturally."

Because many of the students would later join in the work of discovery, the letter made clear that students' main goal should not be to memorize all plants—though students should memorize as many as possible, along with their virtues and uses. Instead, they should learn the plants in such a way that they can carry their knowledge into the field to identify new species and to consider their potential virtues, which could then be tested scientifically.

The letter to the two draftsmen to be hired for the expedition began by explaining what kinds of images they should produce and how: "These Artists must restrict themselves to copying Nature exactly, without any efforts at adornment nor adding anything from their imagination. Therefore they must limit themselves to depicting precisely what the Director, Botanists and Naturalist determine is worthy of depicting [regarding particular specimens], and they must do so under their supervision, obediently listening to whatever they instruct, whether it relates to achieving excellence in the Drawing of the [specimen] or in anything else the Professors deem important for the knowledge and identification of the Plants and Animals, even in cases when it is necessary to depart [from what the artist sees in his specimen] or at times exaggerate proportions."[25]

At first glance, this letter of instruction seems to contradict itself. It ordered that the artists "must restrict themselves to copying nature exactly," but then told them to depart from their model in any way their botanical supervisors might instruct. Thus, even as the draftsmen were ordered to draw nature without deviation, it is clear that what they were being told to draw was not what they had in front of them, but the structure of knowledge imposed on nature by Enlightenment science.

The letter went on to instruct them to draw to the side of the main figure the parts of the flower and fruit, depicting the most important parts of the anatomy to facilitate easy recognition. It expressed a preference for the draftsmen to incorporate color into their images but acknowledged that this might not always be feasible because the goal was to get as much work done in as short a time as possible. And so it advised the draftsmen to use color selectively, only on those parts that, for their brilliant color, uniqueness, or other reasons, really merited it. Usually it would suffice to show in color only one flower, one fruit, and one part of the plant, leaving the rest in "tinta de China" (that is, rendered with black ink lines and wash).

Unlike most of the original members of the expedition, who came from Spain, the draftsmen were to be recruited locally within Mexico. There is no explanation why, but various letters do lay out the exact process by which this was to be done. So important was the role of draftsmen that, before he would authorize the expedition, the king wrote to the viceroy in 1786 to make sure that a process could be devised to identify "two

Draftsmen [*Dibujantes*] who [could] perfectly imitate nature in its plants, flowers, and fruits as desired."[26] Once the viceroy assured him that such talent could be found in New Spain, the king issued his authorization. In 1787, as the rest of the expedition team arrived from Spain, the selection of draftsmen began within Mexico.

The first step was to identify the top four art students in the Academia de San Carlos in Mexico City, as determined by the academy's founder and director Gerónimo Antonio Gil (1732–1798).[27] The Academia had been founded only a few years earlier as part of the same Bourbon Reforms that led to the creation of the Royal Botanical Expedition and the School of Mining. It was originally established as the School of Engraving in 1778 and was housed in the Casa de Moneda (Royal Mint). Gerónimo Antonio Gil of Zamora, Spain, served as the School of Engraving's inaugural director and professor, and took on the important task of overseeing the depiction of the king on Mexican silver coins, which at that time were the most widely circulating objects in the world. The school almost immediately expanded beyond coins into other forms of engraving and then into drawing, painting, and other fine arts. In 1783, it officially converted into the Academia de San Carlos de Arte, under the direction of Gil. Modeled after similar institutions in Rome and Paris, and the Academia de San Fernando in Madrid (founded in 1752 as part of the same effort that led to the creation of the Royal Botanical Garden in Madrid in 1755), San Carlos was the first art academy in the Americas. Initially it relied on a Creole faculty, but in 1786 transitioned mostly to Spanish faculty who had trained in Madrid's Academia de San Fernando. Students at San Carlos studied free of charge, and, following a policy that was unusual for its time, the academy accepted not only wealthy Creole students, but also mestizos and Indians. It even provided stipends to impoverished indigenous students.[28]

The top four art students identified by Gil in 1787 then underwent specialized training in botanical illustration overseen by Cervantes and other members of the expedition, during which the finalists were taught to properly "draw herbs, plants and flowers." After that, the top two were selected. On April 15, 1788, at the conclusion of this process, Vicente de la Cerda (dates unknown, sometimes signed as Vizente de la Cerda) and Atanasio Echeverría (1771–1803) were declared the winners of the competition and hired as members of the Royal Botanical Expedition. Their appointment left an unusually dense paper trail across multiple archives, suggesting that a large number of people were invested in the hiring of competent draftsmen and were eager to know the outcome.[29] Echeverría seems to have stood out since the start, as indicated by a letter from the head of the expedition Martín Sessé y Lacasta to Gómez Ortega, in which he described how, during the training, at "barely 15 years of age," Echeverría "casually completed" impressive illustrations of several plants, plus a butterfly so lifelike "it looks ready to fly off the paper."[30]

It appears that Gil was being pressured by royal officials to reduce the expedition's expenses by depressing the draftsmen's salaries. He explained that he had heard that illustrators for the expedition to Peru received 1,000 pesos per year, with an extra 500 pesos per year while on expedition. Since De la Cerda and Echeverría were locals who did not have to travel from Spain, and the cost of living was perhaps higher in Peru than in Mexico, Gil proposed that they need be paid only 700 pesos per year while in Mexico City and "a little more while on Expedition."[31] The Crown accepted Gil's salary recommendation, but then, citing financial difficulties, further reduced it to 500 pesos and 1,000 pesos while on expedition. This low salary would haunt the expedition throughout its time, to the point that Sessé and other members each ended up contributing part of their own salary to augment the artists' income.[32] Despite their low salary, the draftsmen dedicated themselves to producing superior work.

Kärin Nickelson has pointed out that Linnaean botanists preferred "an aesthetically inferior drawing to" what they might consider "an incorrect one."[33] Therefore, when selecting the top two candidates from among the four put forward by the academy, members of the expedition were likely most concerned with whether the aspiring draftsmen were able to retrain their eye to see with a Linnaean optic, and to convey this through drawings and painted images.

The artists had only received a few months of training in botanical illustration before the final selection. While they had no mentors on hand who had already undergone such training, they almost certainly had access to the textbooks used by students within the garden and school of botany and medicine, which included Linnaeus's tables of the expected shapes of each part of a flower's reproductive organs. The expedition might also have brought along some European botanical illustrations for the students to copy, though there is no record of such items in inventories, nor are they mentioned in descriptions of the training of the draftsmen. Given what the documents do reveal, it seems that the students had to reinvent aspects of the illustration process for themselves, guided only by a textbook and the instructions and corrections offered by Cervantes and other members of the expedition.

By all accounts, De la Cerda and Echeverría learned to produce stunning illustrations, both in terms of complying with Linnaean theory and in terms of aesthetic beauty. They created these works on "Papel de Holanda" (referring to high-quality Dutch paper) with tools and supplies sent from the Royal Botanical Garden in Madrid.[34] When Mociño later showed more than two thousand of the expedition's images to Augustin Pyramus de Candolle (1778–1841), the prominent Swiss botanist praised not just the beauty of the images, but also the artists' skill at depicting the details that botanists needed to figure out how each plant fit within "known families and genera."

While both artists perfectly met the needs of Linnaean botany, de Candolle noted that Echeverría stood out as particularly talented, and even "was superior to the majority of" European botanical illustrators "in the precision of his drawings, and in the beauty and rigor of his coloration."[35] The German naturalist Alexander von Humboldt (1769–1859) was similarly smitten by the quality of Echeverría's illustrations, which he judged "bear a comparison with the most perfect productions of the kind in Europe."[36] De Candolle was so impressed by Echeverría's illustrations, in fact, that he named an entire plant genus after him: the *Echeveria* succulent genus of the stonecrop, or Crassulaceae, family, of which the draftsman had produced a beautiful illustration.[37]

In evaluating these draftsmen's work, it is useful to keep in mind that they were only around fifteen or sixteen years old when they were first appointed to the expedition in 1788. Their youth had the benefit of making them impressionable regarding the logic of the system in which they had to operate. It also meant that they were not yet fully developed as artists when they first started. As a result, a look across the illustrations they created reveals the draftsmen's process of unlearning approaches to art they had learned in San Carlos and their acquisition and gradual refinement of new techniques and ways of seeing that they acquired as part of the expedition.

Two thousand sheets of paper with botanical illustrations produced during the expedition have survived, and these likely represent only a fraction of the total number the draftsmen produced. Of these, none is dated, and only a couple of them are signed. These images are held by the Torner Collection of Sessé and Mociño Biological Illustrations, Hunt Institute for Botanical Documentation, Carnegie Mellon University, Pittsburgh, Pennsylvania, and are available in stable online links as high-resolution images on the institute's website: huntbot.org/torner/. For convenience, I will refer to them by their object numbers and include reproductions of only some of the many drawings to which I refer.

With one exception, the few works that bear the artists' names are relatively crude compared to most of their illustrations, suggesting that these signed items likely were early drawings, perhaps produced as part of the training process. De la Cerda's signed depiction of a member of the flowering vine family Polemoniaceae (item 1436), for example, betrays his training in still life and also reveals that he had not yet mastered how to capture perspective and texture when showing a soft-trumpet flower structure. And the details of the dissected flower, fructification organs, seed, and fruit seem haphazardly arranged in the upper corner.[38] Echeverría's depiction of *Fouquieria* in item 0411 (figure 12) departs more successfully from the style of a still life but comes across as indelicate and unable to convey the texture of the leaves or the flower petals. It appears amateurish compared to how he later painted similar leaves of *Suriana maritima* L., a member of the Simaroubaceae family, in item 1017 (see figure 13, detail). That the plant illustration in 0411 also included a painting of a *Daritis thetis* moth inserted into the lower corner suggests that the image was probably composed early on to evaluate the current state of the artist's skills.

By correlating some of the illustrations with the expedition's itineraries—which other scholars have reconstructed—and by engaging in a stylistic analysis, we can gain a general idea of the order in which they were produced, and by which artist. The works that seem to have been created during the early years of the expedition tend to be cruder than those painted later. The artists appear to have been trying to figure out how to move away from still-life painting toward consistent conformity with Linnaean expectations, and they each did so in their own distinctive manner. It is easier to identify particular images as Echeverría's than to discern which are De la Cerda's because their styles looked similar early on and also were similar when they were doing field drawings or participating in

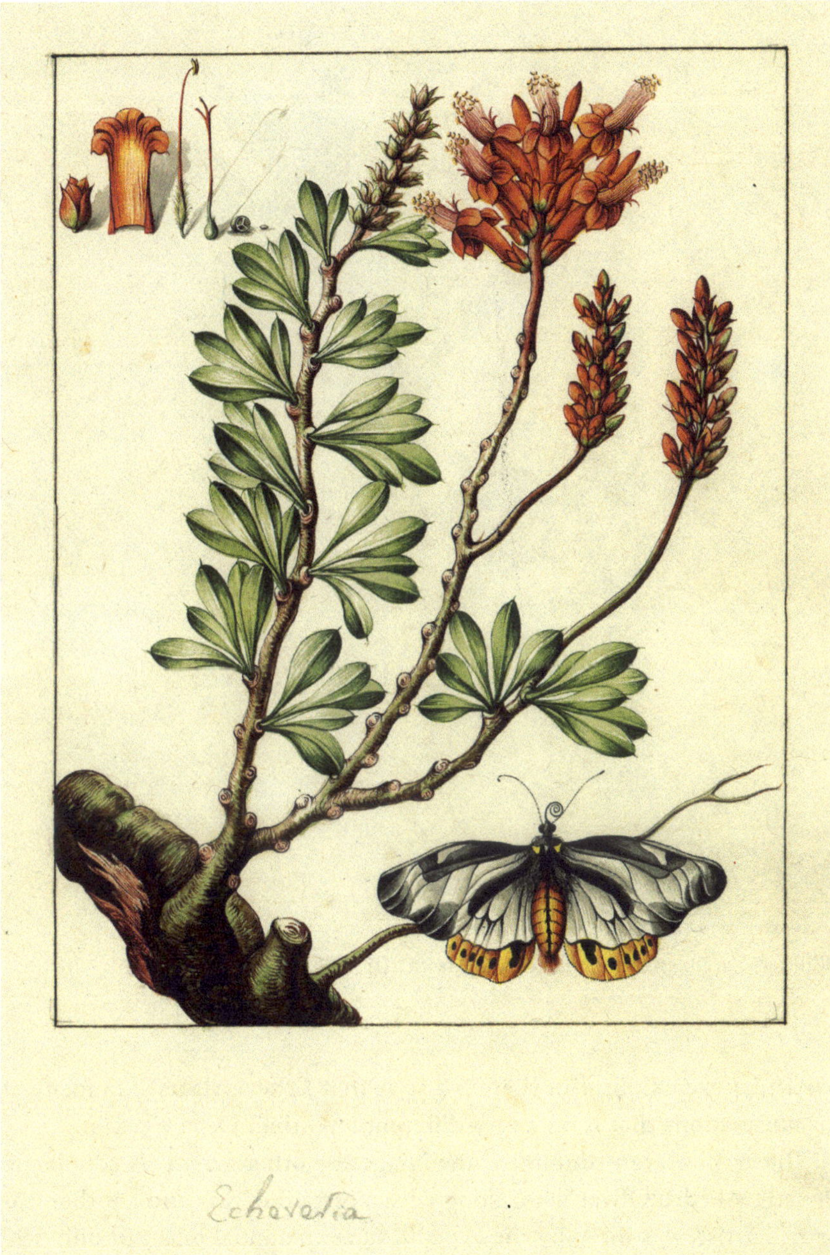

FIGURE 12 *Fouquieria*, Item 0411, Torner Collection of Sessé and Mociño Biological Illustrations, Hunt Institute for Botanical Documentation, (TCHI) Carnegie Mellon University, Pittsburgh, PA.

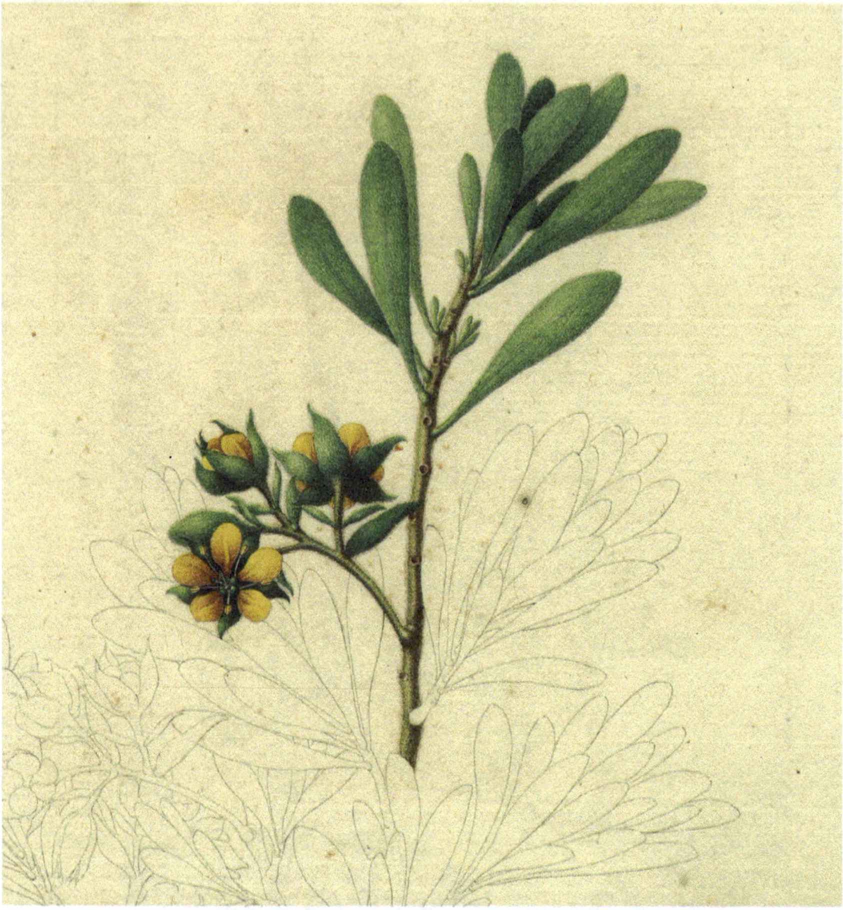

FIGURE 13 *Suriana maritima* L. (detail), item 1017, TCHI.

particular excursions. But there is a style that Echeverría used in many of his illustrations that is on a very different level than De la Cerda's.

The archive even contains some images by other artists. A couple are signed by Pedro Oliver who, along with a certain Alvear and another unnamed artist, was hired by the expedition at an annual salary of only 190 pesos to make copies of the images by Echeverría and De la Cerda and to fill in for them when they were unavailable.[39] Alvear and the unnamed artist are likely responsible for some of the cruder illustrations, such as the clumsily executed drawing in item 0202 (*Smilax rotundifolia*).

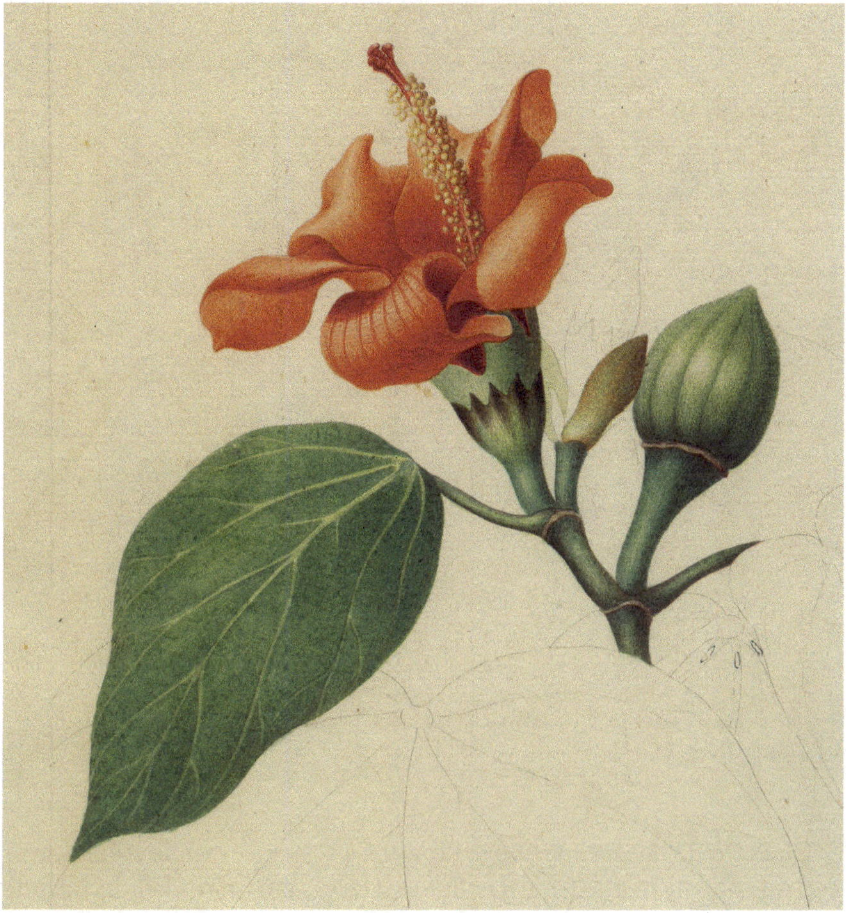

FIGURE 14 *Hibiscus elatus* Sw. (detail), item 1423, TCHI.

As his personal style gradually emerged, Echeverría moved toward delicately inked lines and a painting style that made each plant almost look like it is glowing (see, for example, figure 14, detail of item 1423, originally titled *Azanza insignis*, though later reclassified as a type of Malvaceae called *Hibiscus elatus* Sw.). Echeverría frequently rotated leaves and petals, made strategic use of color, and carefully spaced elements to clarify their shapes and details, as is apparent in figure 19, TCHI 1495 "Apocynaceae," later in the chapter. He also showed particular skill at depicting flowers on the plants in such a way that sometimes made it unnecessary to add inset details.

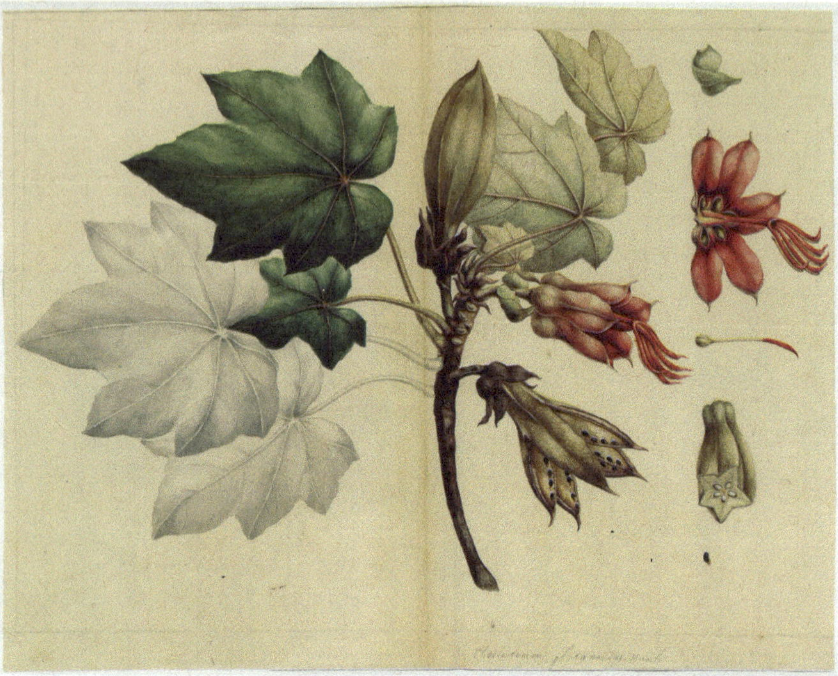

FIGURE 15 Sterculiaceae. *Chiranthodendron pentadactylon Larriategui* (árbol de manitas), item 0172, TCHI.

De la Cerda, meantime, moved toward a style that, while still impressive, was heavily worked, both in line and color, and more likely to include shadows and other features that preserved his early still-life training. He tended to be less selective about which parts of the plant he depicted in color, and more likely to follow the Linnean expectation that he arrange details of the flower and its parts on the top or bottom corner of the page. While De La Cerda's work was less refined than Echeverría's, it could also be quite spectacular (see, for example, his painting of the *árbol de manitas*, item 0172, figure 15).

Some of the artists' mature works can be easily differentiated, but others are impossible to assign authorship to. This might be because of the conditions under which they were painted, the purpose of the particular illustration, or the specific instructions given to the draftsmen by the botanists at that moment. What all the images have in common is that they were created to straddle two goals that were in tension with one another: (1) stripping plants from their context to put them naked into the Linnaean system; and (2) selectively incorporating local information that could be useful for figuring out whether the plant was new to science.

Throughout the years of the expedition, Sessé consistently advocated for better pay for the draftsmen. As he stated in a letter from January 11, 1794: If "at the time I agreed . . . to their salary I had foreseen their hard work and outstanding talent," then "I would have insisted that they be given the same salary as the draftsmen of other Expeditions, that is, one thousand pesos." Now that they have demonstrated that "they are not just equal but superior" to those other draftsmen, their salary should be increased.[40] What made these draftsmen particularly valuable was their demonstrated ability to describe that which had never before been described according to the standards of Enlightenment science, and to navigate each step of this process from the first sketch hastily executed in the field to the final product ready to be exported to Madrid. It is to their process that we now turn.

STRATEGIES TO DESCRIBE THE NEW

The botanists and draftsmen of the expedition marched on the front lines of the Enlightenment project, wrenching plants from their local context and thrusting them into the alien Linnaean system. As they traveled across the breadth and depth of the colony, they rarely indicated the location where they collected a specimen and made no observations about the range in which a species seemed to thrive. All that mattered was that the plants were from New Spain. The only exception was plants that came from the northern outpost Nutka, where Spain and England dueled over the claim to ownership. But even these were not marked consistently. And illustrations from Nutka were sometimes mixed with those from other locations.

While this lack of interest in location or range might seem odd to us today, and certainly has proven frustrating to later botanists, it was so commonsensical to the members of the expedition that they never even commented on the practice. As Linnaeus colorfully explained in 1737: "Native locality does not make a specific difference. Locality neither canonises nor changes anyone; as the proverb goes, not even a pig is changed by being taken to Rome."[41] But it was not so simple, especially for members of an expedition designed to assert the empire's power over its colony by identifying new plants on the botanical frontier.[42] The system they employed was universal in its ambition, but it did not obliterate the importance of place as thoroughly as Linnaeus might have wished. Instead, it changed how place was conceived.

The effort to erase local information did not occur all at once. Instead, it unfolded in calibrated steps. This is demonstrated by the fact that, during the early stages of their work, members of the expedition sometimes incorporated certain kinds of information about local place, cultural practices, and environment into the text and images. One way they did this was by adding captions that explained indigenous uses of particular plants. The plants most likely to carry such information were those that the expedition marked with some variant of "Hrz," indicating that expedition members thought the plant had been described by Hernández in the 1570s. Of the 2,000 items (sheets of paper) in the Torner collection, 1,792 contain plant illustrations. Most of these have no writing on them other than numbers for temporary correlation during field work, but some contain written notes. Of the Torner items showing plants, 188 (10 percent) are marked as Hrz. Twenty-six of these Hrz items contain information about the plant's uses, written in Latin. Item 1091, for example, depicts maguey, and states in Latin "Hardly any plant is more useful either in medicine or in economy, although today they only make [from it] the wine they call Pulque." Item 0034 shows a kind of spiderwort, and states "The root is resolutive, astringent, and aphrodisiac. The flower gives off an excellent blue color." It is possible that this information is derived from Hernández. But there are an additional fourteen items with this kind of information that have no Hrz designation and therefore would not have derived their information from Hernández. Item 0627, for example, shows a kind of mimosa and states "The fruits, warm and like aphrodisiacs, are eagerly eaten by the Indians."[43] And item 1123 shows a kind of sedum and states "It relieves pain and inflammation, and heals fresh wounds." If this information did not come from Hernández, then perhaps the members of the expedition acquired it through conversation with local people.

Considering the artistic style of the images that contain such notes, it is likely that all these were done in the early years of the expedition. None of the illustrations done later contain this sort of text. This does not mean that such information was no longer collected. It probably would have been recorded in other ways, such as separate field notes that could later be discarded, thus erasing the botanists' reliance on information that came from outside the Enlightenment scientific structure. This would be consistent with the exploitation, then erasure, of indigenous knowledge described in the previous chapter in relation to the botanical garden.

The images also betray other intrusions that were supposed to be eliminated by the Linnaean system. One of the most obvious is the perpetuation of Náhuatl nomenclature, which appears in ninety-three of the items in the existing collection, despite Cervantes's 1788 admonition that such names had no place in modern science (see chapter 3). Some of these incidences can be explained by the expedition members' efforts to correlate their work with Hernández's (keeping in mind that the Náhuatl name assigned by Hernández was likely constructed as part of his system based on Náhuatl, and not necessarily the vernacular name used by local people; see chapter 2). But an additional thirty-one of the items with a Náhuatl name have no Hrz designation. This raises questions about the source of the name. Most of these are marked "Mex," indicating that the name came from "Mexicano," as Náhuatl was often called at that time. We see this, for example, with Quauhaiohutztli, Papaloquelite, Tlatlacoxochitl, and Pipiloxihuitl (items 0259, 0266, 0269, and 0570), some of which also carry a separate Linnaean Latinate binomial, indicated by "Linn." Others carry the indigenous names less formally. We see this, for example, with item 0627, which provides the Náhuatl name "guaxe," labeled as "vulg," meaning it was a vulgar, that is, a colloquial, name.

Sometimes something even more intriguing occurs regarding images and texts. Item 1928 shows details of fruits that look similar but that come from different plants. Each of the four fruits is accompanied by text explaining its use. One, for example, explains that the image is of "Fruit of the plant of 20 pieces. It is called Tehuacan Tiliapas and, when it is unripe, Tempestis. This they eat unripened in place of olives, and ripe, in place of *fruta regalada* [likely referring to zapote]." Analysis of the plants' very dissimilar leaves, growth habit, and flowers would easily have distinguished them from one another. The descriptions of the fruits, therefore, was related to an interest in local uses rather than to the botanists' need to make Linnaean taxonomical determinations.

When attempting to make sense of plants that may never before have been scientifically described using Linnaean standards, members of the expedition probably considered it unwise to eliminate local information too early. It took time to figure out what information was useful and what was not, just as it took botanical illustrators time to figure out which parts of a specimen should be recognized as characteristic of the type and which should be dismissed as noise. Without a dense structure of other illustrations of the same species that could be copied and perfected, the expedition's botanical illustrators had to rely on other ways of figuring things out. They had to be particularly steeped in Linnaean theory and deeply versed in acceptable variations, and they also had to acknowledge, even if only provisionally, aspects of local culture and environment that were outside the Linnaean system. This led to practices on the part of the expeditions' draftsmen that later would need to be erased, like footprints swept out of the sand after passing.

Historian of science Mauricio Nieto reminds us that the "traditional notion of 'discovery' suggests an individual who, in a moment of luck and cleverness while walking through the forests of the Americas," simply comes across things previously unknown, so that he can fancy himself the independent discoverer. But the actual act of discovery "of a new genus or species, far from being an event occurring at a specific time and place, must be understood as a prolonged process of elaboration in which the incomparable is made comparable, the alien is domesticated, and the strange is rendered familiar and conquered."[44]

When members of the expedition suspected that a plant might be new, they often, but not always, indicated this by writing "N" on the page. This probably was to remind themselves to look more deeply into the matter once they returned to Mexico City, or to invite others to investigate the possibility once they were able to circulate their information through European networks.[45]

To aid this process of discovery, botanical illustrators in the expedition also depicted details that would normally be erased by Linnaean science, such as the plant's roots. Though standard practice in Linnaean illustration was to exclude plants' roots, 42 percent of the items in the Torner collection that show plants include the roots. In the search to identify which species were new to science, and how they should be categorized in relation to other plants, it was perhaps wise to capture any information

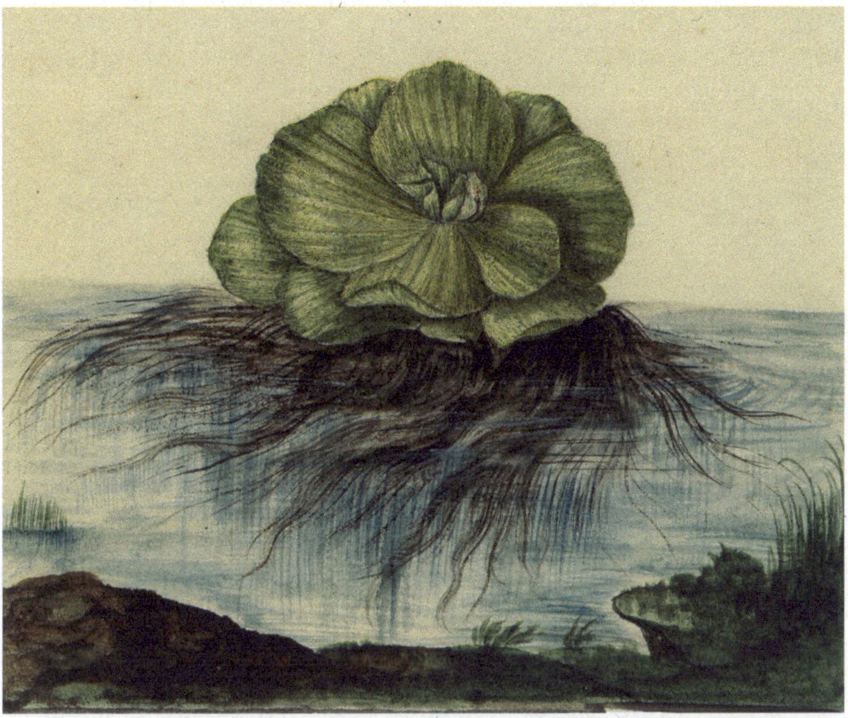

FIGURE 16 Araceae. *Pistia stratiotes* L. (detail), labeled "Pistia trapoides?," item 0919, TCHI.

that might provide clues. This same principle seemed to guide the draftsmen in their inclusion of information about the local environment.

Many of the expedition's illustrations carry a surprising amount of information regarding the plant's local ecology. Item 0509, for example, shows *Hydrocotyle racemosa* DC (a member of the Umbelliferae family) growing with its roots floating in clear water. Items 0693 and 0709 show plants growing in water, but make it apparent that the water is shallow, with the roots possibly extending down into the muddy bottom. Item 0207 (figure 23, later in the chapter) shows a plant growing in moving water, with the roots anchoring it in place as the current pulls along the plant's upper flowering portion. It also shows the muddy shore. Item 0919, labeled as "Pistia trapoides?" goes even further (see figure 16, detail), showing not just the shoreline of the pond where the plant is growing, but also some of the particular plants and grasses that inhabit the water's edge,

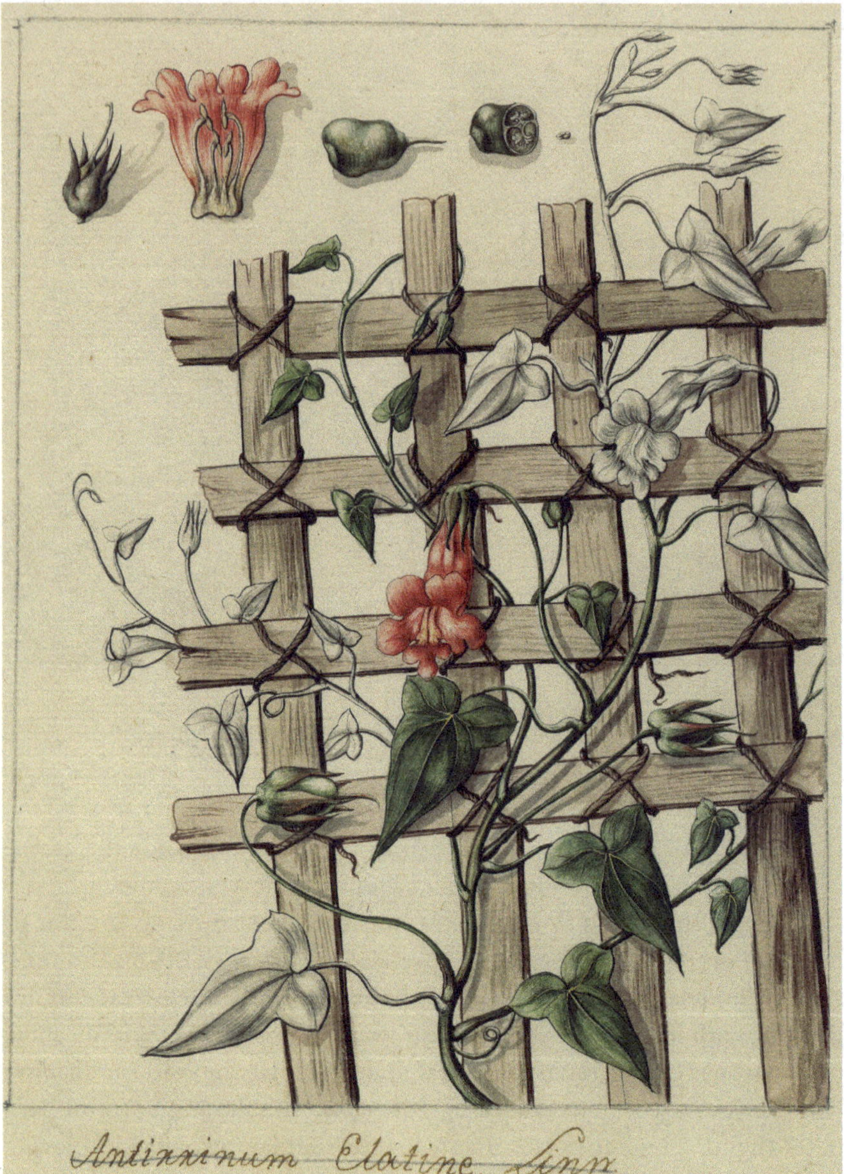

FIGURE 17 Scrophulariaceae. *Maurandya*, tentatively labeled "Antirrinum elatine Linn" (note that the vine is climbing a trellis), item 1637, TCHI.

almost like a painting of a miniature landscape that offers hints about the plant's ecosystem. Given the uncertainty of the plant's identity, as indicated by the question mark, the artist perhaps hoped that these inclusions might help botanists place the plant within the Linnaean system. Items 1128 and 1681 show plants growing out of rocky ledges. And item 0976 includes some of the smaller plants and grasses that might accompany the plant on its rocky ledge. Echeverría and De la Cerda typically displayed vines climbing generic twigs or branches, thus conforming with the goal of showing plants free of context. But item 1637 (figure 17), which was tentatively labeled as "Antirrinum elatine Linn," shows a vine climbing a man-made trellis, revealing it as a garden plant growing in some sort of domesticated space. Each of these hints conveys information about the context in which a plant grows, seemingly as clues to assist botanists in the laboratory or in distant armchairs.

Illustrations also sometimes included a sense of the passage of time, another violation of Linnaean protocol. A species of Rosaceae shown in item 1503 (see figure 18), includes three leaves that display insect damage. Two seem freshly munched on, while a third looks like it was eaten some time earlier and has healed, thus introducing both idiosyncrasy and the flow of time. We see this sense of time also with the plant's flowers, some of which seem to have dropped, leaving only their empty hypanthia, or receptacles, while others are in full bloom. Both of these details—the healing leaves and aging flowers—betray a sense of the specific and of the passage of time, two things that Linnaean draftsmen typically sought to exclude.

The illustrators and botanists of the expedition were too experienced and closely supervised for this to be accidental. Instead, it appears to be an intentional incorporation of ordinarily proscribed details to aid the expedition's efforts to identify the particular plants. Only once members of the expedition had identified a plant with some measure of confidence, however tentatively, could they cleanse its depiction of evidence of this process so that it could be displayed as a perfect type, outside time and place.

Several of the illustrations even convey information about the plants' relationship to insects. Some of these seem to be plants in whose identity the botanists probably were confident. Item 0159, for example, shows a moth infestation in a particular kind of *Quercus*, while 0264 shows a cochineal infestation on a kind of nopal. Others seem to be plants of whose

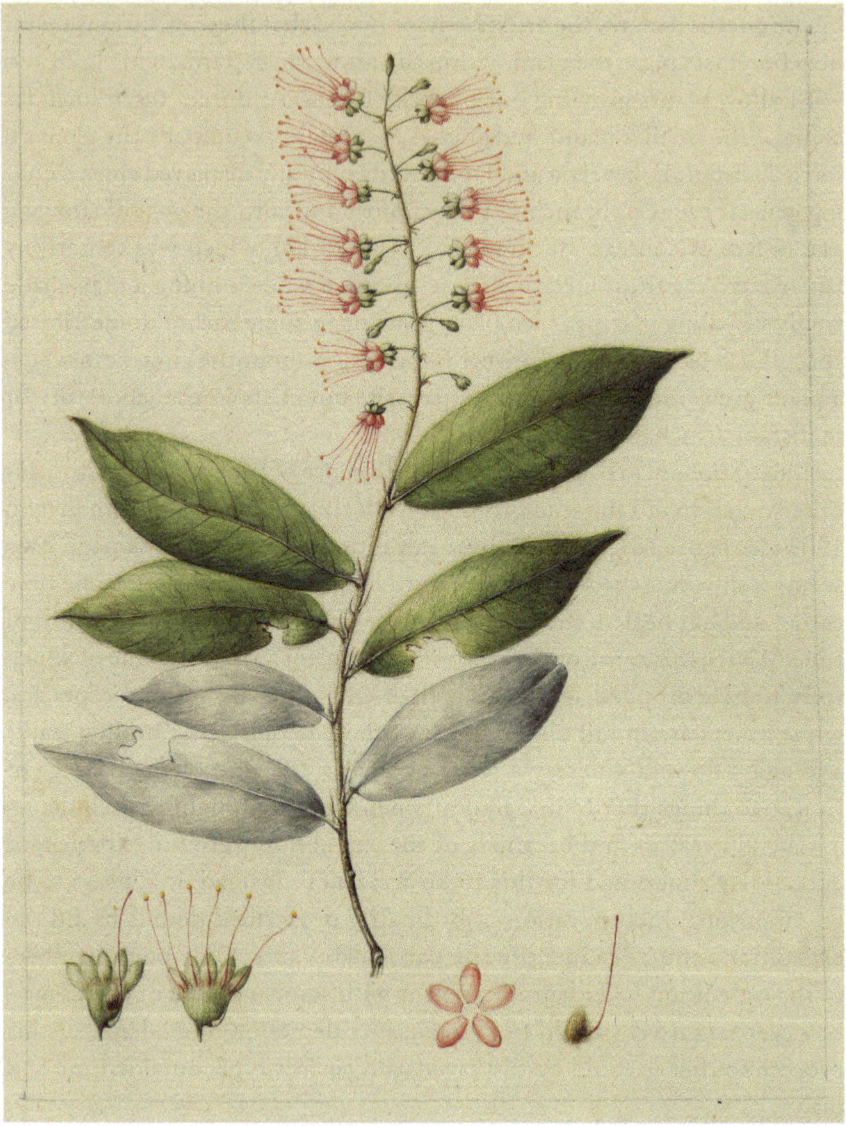

FIGURE 18 Rosaceae. *Hirtella oblongilgolia* DC, item 0503, TCHI.

identity the botanists were perhaps uncertain. In item 0823, one of the colored leaves of a type of hibiscus shows in detail the patterns of caterpillar damage, perhaps to gain some insight about the plant as indicated by its relationship with certain kinds of insects. Similarly, item 1444 shows damage on two leaves: one in black and white to highlight the pattern of damage, and one in color, revealing how the pattern of damage relates to the leaf's veination.

We see this taken even further in item 1495 (see figure 19) which shows a species of Apocynaceae, tentatively labeled as "Plumieria multiflora." Here, Echeverría carefully complies with Linnaean practice by lightly sketching much of the specimen in black ink, while painting in color only the parts that are most essential to its identification, including a few flowers at different stages of florescence, and two leaves, one showing its top and one showing its bottom. But here we get the surprising inclusion of a black, yellow, and red caterpillar eating the edge of one of the leaves. The level of finish in this illustration makes clear that this was a carefully considered inclusion. We might speculate that perhaps this caterpillar preys on this particular species, and so maybe Echeverría included it to help botanists distinguish this species from other similar species, such as the Apocynaceae tentatively labeled "Plumieria emarginata" in item 1032. While there is no written explanation, it is clear that this violation of the Linnaean expectation that a plant species be isolated from its context is intentional and that it seems to have been done as part of the process of exploring the botanical frontier.

In addition to expunging references to local places or cultures, Linnaean scientists strove to minimize the human presence. In his early Renaissance-era study of New World nature, Oviedo had emphasized his own role as eyewitness. Linnaean botanists and draftsmen, by contrast, strove to erase outward evidence of their personal role so that the things they described and depicted came across as disconnected from the particularities of personality or circumstances. One of the reasons that the previously mentioned item 1637 (see figure 17, above), with its depiction of a vine climbing a man-made trellis, is unusual is that it invites the viewer to contemplate the human story surrounding the plant. On other occasions the human presence asserts itself even more explicitly. In item 0064, for example, below the botanical illustration, there is a faint sketch of a woman who appears to be in prayer. Perhaps the draftsman, calling on his

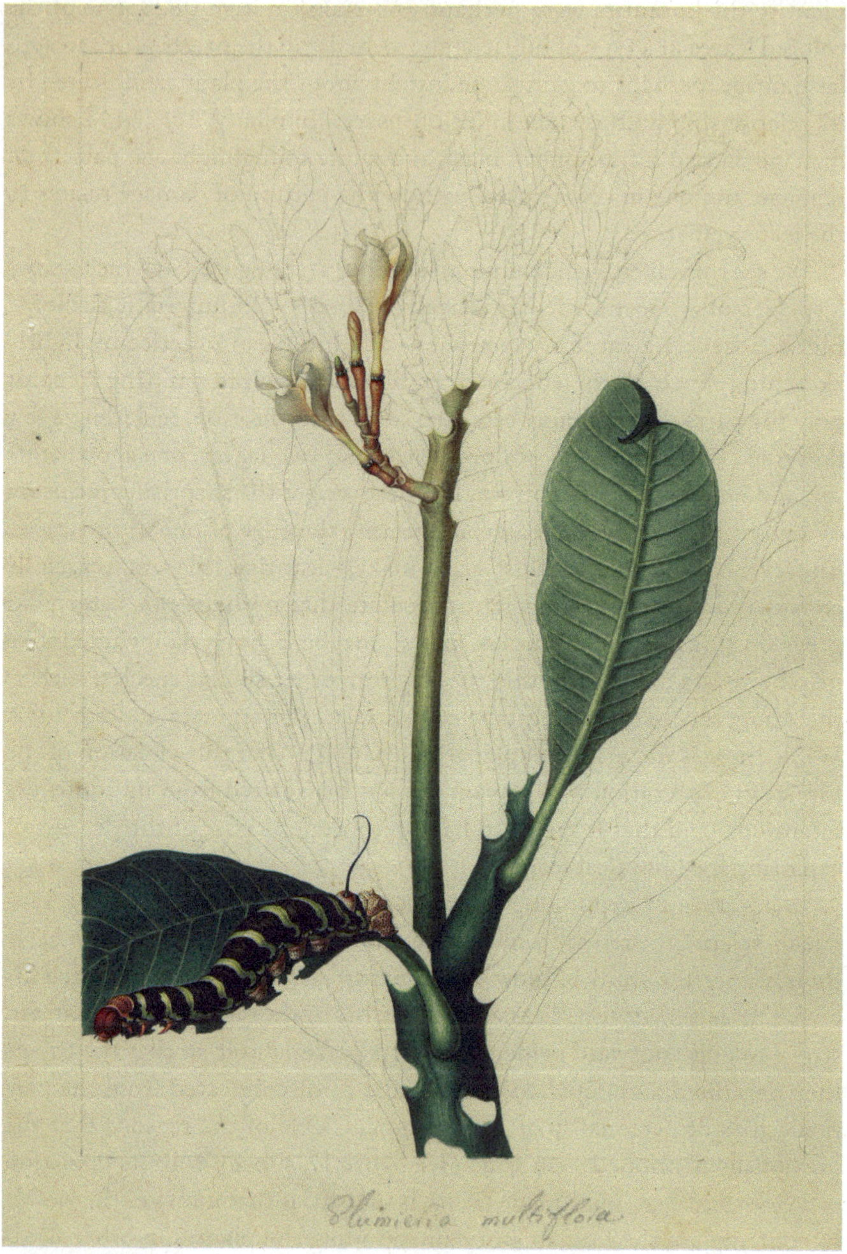

FIGURE 19 Apocynaceae, "Plumieria multiflora," item 1495, TCHI.

previous academic training, was inspired by his observation of a local person, or else copied a detail from an altar painting. We see this kind of intrusion again in item 1917, which contains below the botanical illustration a small light pencil sketch of the head of a man, possibly a member of the expedition (see figure 20, detail), who is wearing a collared shirt and a cap of some kind. Unlike some of the other intrusions, which are clearly intentional, the woman in prayer and the man's head seem like unsanctioned slippages that reveal the life of the artist and of the people around him.

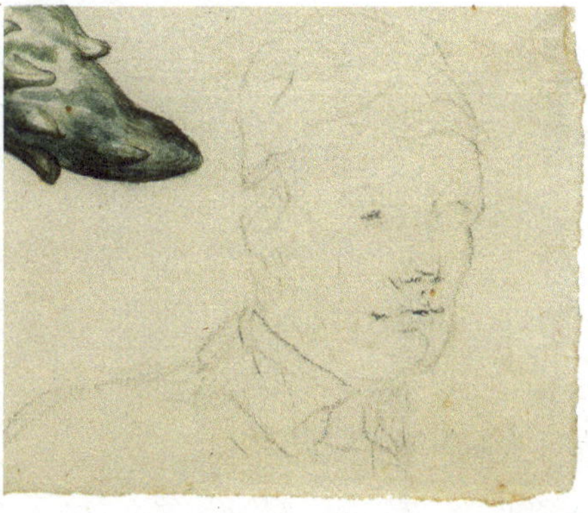

FIGURE 20 Drawing of a man's head, detail of bottom-right corner of item 1917, TCHI.

Unlike these rare unsanctioned slippages, most of the unusual attributes we see in the expedition's botanical illustrations seem to have been part of the process of trying to identify and describe novel plants. It is clear that these botanical images were not meant to be final statements but rather tools for identification, so that a taxonomic decision could be made later. We also know that there was a plan to redo the images into finished versions after the completion of the expedition. In 1805, after the end of the expedition, Echeverría even drew a mockup of *Dumerila procumbens* (item 0422, see figure 21) while in Madrid as a model for what these finished images would look like for final publication.

Many of the items in the Torner collection compare multiple images, side-by-side, to aid expedition members in distinguishing between plants that might belong to different taxa. Item 0645, for instance, compares two very similar plants that both are marked with "N," arranged to help figure out whether they are of the same species. If they are determined to be the same species, then the indicated slight differences in flower and leafing pattern should be dismissed as incidental, but if they are determined to be two different species, then these subtle differences in flower and leaf pattern might be indicative of each's type.

Another way they tried to resolve uncertainty was to draw a plant twice, at two different stages of its growth, rather than collapse multiple

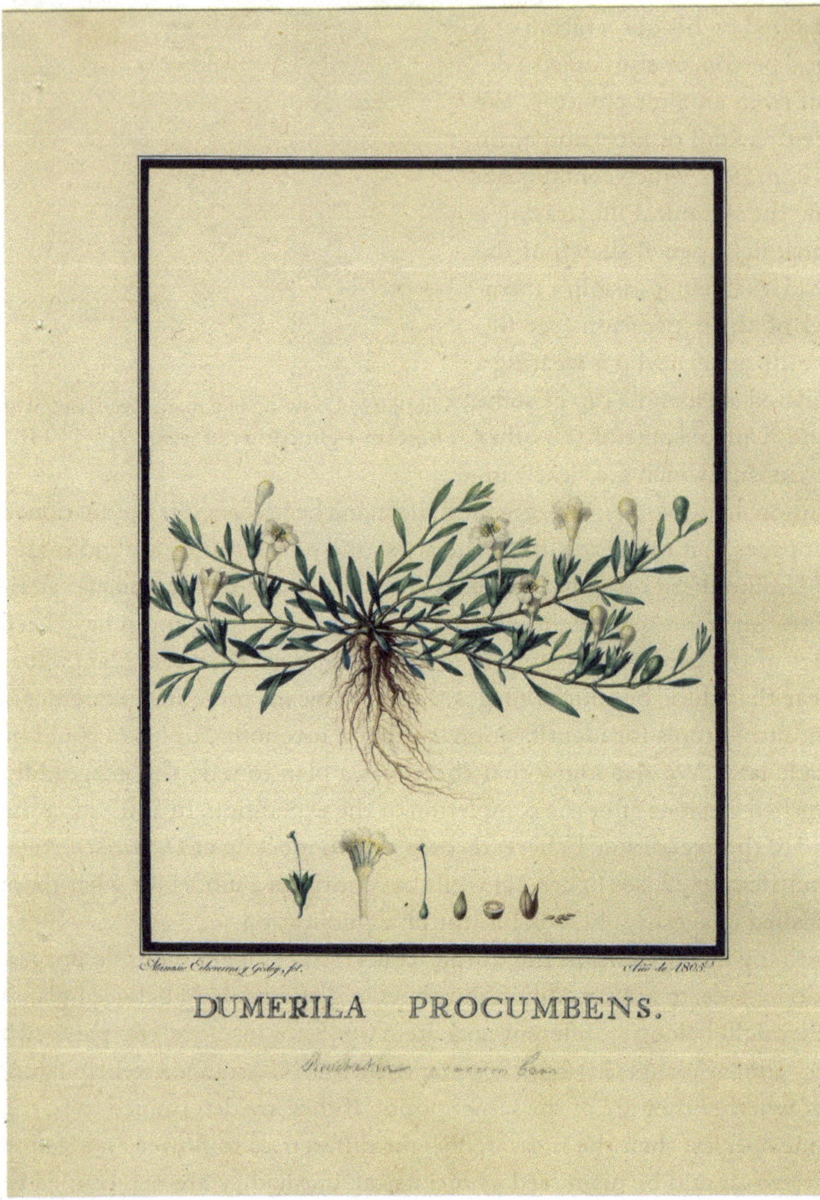

FIGURE 21 Solanaceae. prob. *Bouchetia* (*Dumerila procumbens*), painted in 1805, item 0422, TCHI.

moments into a single image, as was typically done in finished Linnaean illustrations. We see this in item 1934, which shows a plant later identified as *Vibilia glabriflora*. The specimen on the left is shown with the flower in early bloom with the leaves still tightly wrapped around the stem, and to the right the plant is drawn with its flower fully opened and its leaves unfurled. It also is possible that at the time he drew these two images, the draftsman might have been trying to puzzle out whether two specimens were of the same taxon at two stages of growth, or else belonged to two separate taxa.

Other times the inclusion of multiple images might have been specifically for the draftsman's own use, created as a reference tool to guide him as he encountered more plants or as he converted rough field sketches into polished illustrations. Item 1877 is a good example. It contains numerous pencil sketches, some of which include a thin paint wash to convey a rough idea of the color. Scattered across the page also are numerous stray marks from pencils, some of which seem to be the remains of an effort by the artist to give his pencil tip a sharp point. There also are stray brush marks that seem to be to test a color blend before applying it to some other image. We see similar marks on items 1889, 1936, and 1937, alongside finely rendered details of the budding process of various flowers. And item 1921 shows delicately executed details of a member of the Burseraceae family, but they are chaotically scattered across the page, and on the back of the sheet is a large sketch of a seed capsule from some other unnamed plant, along with various notes that do not seem to be related to the drawing. Numerous items similarly include multiple, sometimes several dozen, images on a single sheet of paper, rendered to different levels of finish. These seem to serve a number of different purposes, including as reminders for use in the later production of a more finished image, comparisons to guide the artists' eyes in the field, or visual aids to help botanists later figure out how to categorize particular plants. To lend order to the numerous images on some of these sheets of paper, the draftsmen sometimes drew frames around groups of drawings, or traced lines to link a plant on one end of the page with a flower on the other end (see for example item 1909, figure 22). Item 1901 even includes a detail of a bird head alongside drawings of plants, reminding us that while the emphasis of the expedition was on plants, members also took an interest in animals.

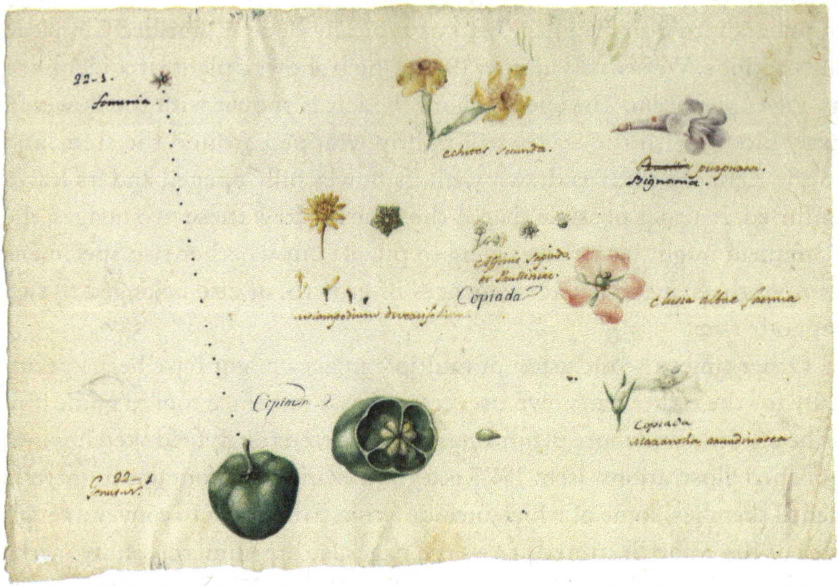

FIGURE 22 Item 1909, TCHI.

Many items included scribbled notes. On item 1567, for example, the draftsman wrote "*sacados en limpio* (figured out, or resolved)," suggesting that he or someone else had resolved some uncertainty related to the plants in the image. And on item 1568 is a scribbled field note in which the draftsman reminds himself to "Comply with the Description [from the botanist] when finalizing [depiction of] the dissection of the berr[ies] and flowers."[46] And some items, such as 1973, 1999, and 2000, show numerous black-and-white sketches of flowers accompanied by detailed notes that recall Linnaeus's long names, such as the previously mentioned long name for sweet vernal grass.

Other times images appear to have been abandoned before adding any of the essential details. We see this, for example with a cactus shown in item 0969, and lightly sketched plants on items 1016 and 1058. None of these contains enough information to help the artist finish the image later in the studio, or to guide a botanist's efforts at categorization. Instead, it seems that, shortly after the artist started his sketch, the botanist overseeing him perhaps decided the image was unnecessary.

Items such as these remind us that almost all the botanical illustrations created by the expedition were, on some level, tools for the process of discovery rather than final images. At the same time, the draftsmen likely recognized that many of the images that they painted were destined to become the type specimens for defining new species that the expedition discovered.

Without a paper trail of previous illustrations of each species to guide them, De la Cerda and Echeverría, along with their supervising botanists, had to make decisions quickly in the field, trying to predict which features their contemporaries in Europe might find most important. De la Cerda and Echeverría also had to become extraordinarily steeped in the literature, descriptions of known plants, and Linnaean theory so that they could work nimbly, making difficult choices about when an observed deviation from expectations was to be ignored or emphasized.[47] Draftsmen in Europe working with familiar species could strive to perfect their depiction using the long-name of the species and previous depictions. But draftsmen working in the Royal Botanical Expedition to New Spain, who were charged to draw previously unknown species, had no firm way to be certain whether they were working with a new species or a known one.

When confronted with a specimen about whose status the artist was uncertain, Linnaean training told him that the four most important things to depict were the number, shape, position, and proportion of the plant's reproductive organs.[48] Upon later examining many of the images from the expedition, the Swiss botanist Augustin Pyramus de Candolle complained about how difficult it was to make sense of the expedition's written plant descriptions because they often made reference to field notes that no longer existed, or the expedition members sometimes took incomplete notes on a plant they wrongfully presumed was a known species. But where texts fell short, "the drawings made in Mexico," he explained, thanks to their careful depiction of the appropriate botanical details, "have facilitated my work."[49]

ERASING THE PROCESS

In 1629, a century and a half before the Royal Botanical Expedition, the English apothecary Thomas Johnson had noted:

> Almost every day in the herb market one or other of the druggists, to the great peril of their patients, lays himself open to the mockery of the women who deal in roots. These women know only too well the unskilled, and thrust upon them brazenly what they please . . . Is not the fate of the patients who rely upon the help of such doctors and druggists pitiable? For the doctor relies upon the druggist and the druggist on a greedy and dirty old woman with the audacity and the capacity to impose anything on him. So, it often happens that the patients' safety depends on the herbal knowledge of an ignorant and crafty woman.[50]

Even though Johnson derided the woman as ignorant, it is clear that his fear lay precisely in her *lack* of ignorance, and how she might leverage this advantage against the men of science who should be the heroes in his view of the world. Similarly, Mexican pharmacists and doctors, along with the members of the Royal Botanical Expedition, relied on knowledge from the country's indigenous people. But whereas Johnson's druggists were defenseless against the deceptions of their informants, members of the Royal Botanical Expedition came armed with Linnaean tools of observation, description, and illustration.

These tools of modern science, they believed, empowered them to wade through accounts by local informants to figure out the truths of nature. They could plug plants into defined slots in the Linnaean system to confidently identify specimens and to make progress in their study of the medicinal and economic potential of each species. In the end, they aspired to eradicate any further reliance on local knowledge by wresting control from commoners and placing it, instead, in the hands of elite men of science.

According to Vicente Cervantes, thanks to Enlightenment science "the true naturalist," will "know better the differences that separate certain objects from others so as not to confuse species; the medical researcher will observe scrupulously the qualities and virtues of each [simple] so as to prescribe it without doubt or error; and the able chemist will discover through exact analysis the principles that comprise them [i.e., simples] so that, examining the particular nature of each, might apply each to the uses to which it is best suited." Once the "true naturalist" could classify the plant by class, order, and genus, he would gain full control over it, whether to exploit it for its medicinal attributes or acclimate it to other parts of the empire. Most importantly, he would no longer be vulnerable to the whims of native informants.[51]

Upon completion of their expedition in 1803, Sessé and Mociño spoke publicly about some of the challenges the expedition had faced as it sought to identify plants on the botanical frontier. "So that any imperfections in our images are not blamed on negligence we want to warn that in many cases it was not possible to acquire a complete sample. Of some plants we knew only their fruits, others their flowers, and sometimes only their trunks. We will correct these defects should some future opportunity arise for us to examine these same plants." And, when we "notice in our images a discrepancy," typically in the form of some detail that would pass unnoticed to the untrained, "we indicate so immediately" so that no errors will be reproduced by others. The next step was to take their texts and images to Madrid, where Sessé, Mociño, and Echeverría would collaborate with other "modern authors" to assemble and circulate their findings for the benefit of the Spanish Empire and "universal" knowledge.[52]

Sessé and Mociño explained that members of the expedition sought to correlate their findings with those from Hernández and other scientists, but they explicitly avoided correlating them to anything derived from current Indians, whom they described as "barbarous" and ignorant even of "the things they use most often."[53] To speak otherwise of indigenous knowledge would have risked unmasking as fiction the claim that Enlightenment science did not depend on local people or alternative structures of knowing. Step by step, as the expedition moved its findings from the field to Mexico City, it gradually erased the evidence of this dependency. Now, as the findings were prepared for transfer to Madrid where they would be assembled as a definitive Enlightenment-era flora, the texts and the images would be purged of the final traces of their dependency on indigenous people or local ecosystems. By the end of the process, evidence of Enlightenment botany's predatory dependence on local knowledge would be erased, so that each plant could circulate through metropolitan centers, controlled by imperial agents and European men of science and free of other claimants.

Images played a central role in this process. When something is shown visually, it takes on an aura of truth beyond what can be captured by text alone. Images are arguments, and they wield an irresistible power to convince viewers about the truth of their claims. Because of this, once scientists had mined local information enough to name and claim a plant, they began the process of eliminating from the images any information not sanctioned as part of the Linnaean system. This involved a slow process of refinement that is evident if we compare early images from the expedition to those produced late in the process. For example, in images of water plants drawn early during the expedition, such as *Jussieua repens* Linn., item 0207 (figure 23), the artists included clues about the local environment, such as the kind of water and shore. But later drawings, such as the *Heteranthera obtusa* shown in item 1656 (figure 24), or the *Heteranthera uniflora* in item 1738, eliminated the water, the soil, and other traces of the local environment. As the artists grew more confident, they also stopped

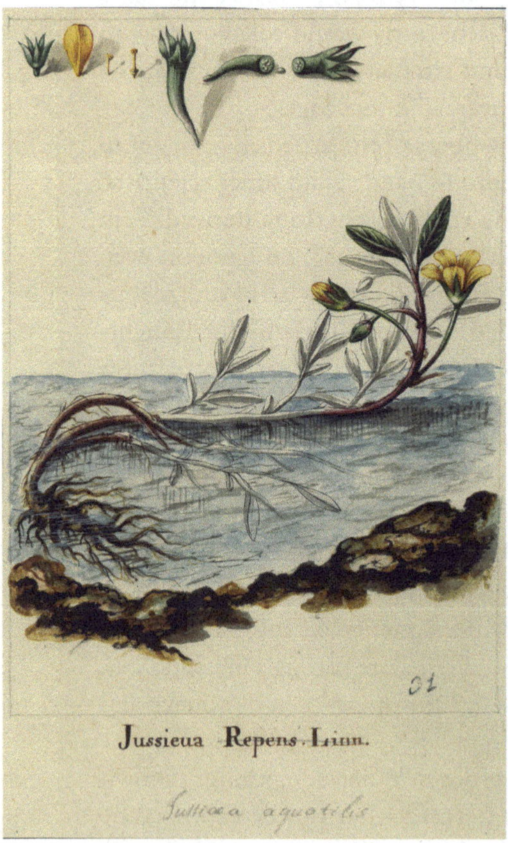

FIGURE 23 *Jussieua repens* Linn., item 0207, TCHI.

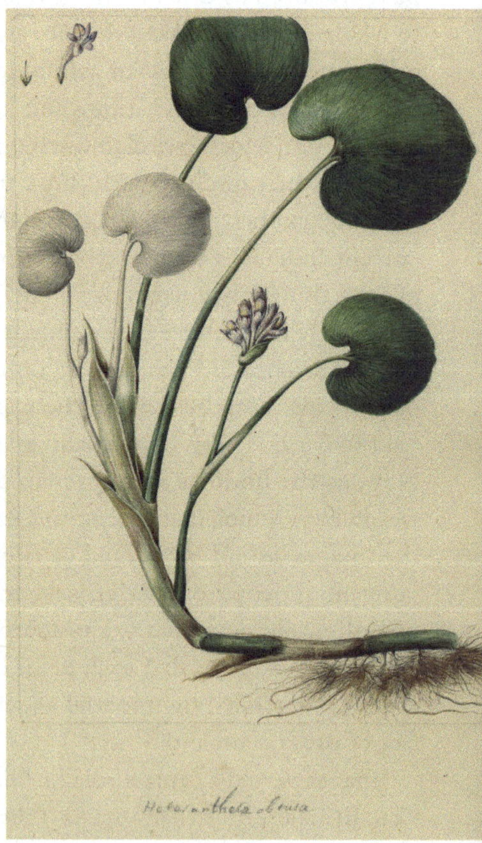

FIGURE 24 *Heteranthera obtusa*, item 1656, TCHI.

incorporating indigenous names and knowledge derived from vernacular culture, as well as information about the local ecological and cultural context. The ultimate goal was to refine the images until all that would remain would be the isolated plant and its flowers, naked on a blank background, ready to be named, claimed, and exploited following the Linnaean model.

CONCLUSION

Hernández in the sixteenth century had esteemed his native draftsmen's ability to improvise to create a novel system for depicting and understanding the plants of Mexico. Sessé, Cervantes, Mociño, and the rest of their team of naturalists equally appreciated the skills of their draftsmen. But what they esteemed about Echeverría and De la Cerda was their ability to conform to the Linnaean optic emanating from the metropole, while subtly incorporating local information that might help botanists slot the plant into the Linnaean system, and to do so in such a way that this local information could later be hidden from sight. As previously mentioned, they then brushed out of the sand the footprints that betrayed their path, until such traces became imperceptible and forgotten. But their tracks could not be completely erased. The compressions remained for those who might seek to recover them.

As for the emergence and deployment of the Linnaean optic as a tool for visualizing nature, there was not one universal outcome. In Britain, Germany, and France, the art of botanical illustration opened botany up to a broad segment of society, as members of the middle and upper classes were invited to appreciate these images and even to try their hand at it. In Mexico, by contrast, botanical illustration gave scientists access to plants from across the colony, but also narrowed public participation because Linnaean illustrations were based on esoteric knowledge rooted in distant metropolitan centers. It widened the distance between common views of plants (whether derived from the Mexican homegrown Enlightenment, Nahua science, or folk practice) and the ways those same plants were known and studied by scientific elites trained in the Linnaean system.

Claims to knowledge about the plants that were rooted in local soils were taken away from the people who lived with those plants daily. The Enlightenment structure for defining what it meant to know nature rendered Mexican plants illegible to those who had provided the expedition with the information it needed, thus making it possible for Cervantes to confidently declare that indigenous people were ignorant even of "the things they use most often." In the coming years, a new generation of Mexican botanists trained in the garden and its school of botany and medicine would step in as the claimants to native resources and knowledge. They would go on to accelerate the process of transforming the conquered landscape into a place whose natural resources were increasingly rendered alien to, and alienable from, the people who had known them most intimately.

CHAPTER 5

FROM EMPIRE TO REPUBLIC

In 1800, the new viceroy tried to cut costs by forcing Vicente Cervantes, director of the Royal Botanical Garden and of the school of botany and medicine, to return to Spain. Back in 1788, Creoles might have cheered the viceroy on, but by the end of the century, Cervantes had become sufficiently embedded within Mexican society that locals protested the proposal. They helped Cervantes win permission to stay until the end of the expedition in 1803, and when that date arrived they successfully appealed for him to stay in Mexico permanently.[1] The school's graduates, meantime, had grown from a small cohort of fewer than a dozen to almost eighty per term at the end of the century. By the time of the independence wars (1810–1821), the garden and its school of botany and medicine no longer felt like imperial impositions. On the contrary, many of the scientists, physicians, and intellectuals whom Sessé and Cervantes had trained now rose to positions of political and economic power, and they incorporated the garden and its school into their own modernizing vision for Mexico. This chapter follows the fate of the expedition after it completed its Mexican fieldwork in 1803 and returned to Europe. It then traces the process within Mexico by which the garden was recast from an agent of empire into a tool for the creation of the Mexican Republic.

FAILURE REDUX

At the official conclusion of the expedition in 1803, Sessé, his lead assistant José Mariano Mociño, his best draftsman Atanasio Echeverría, and several other members of the expedition departed for Madrid to present their findings. Meantime, Cervantes, López, De la Cerda, and others stayed behind to manage the garden and the school. Sessé and Mociño brought with them twenty-seven boxes of manuscripts, a large collection of living plants, a vast herbarium of twenty thousand plant specimens and seeds, and over two thousand botanical illustrations. Their intention was to incorporate this material into the Royal Botanical Garden in Madrid, and eventually to reexport plants and medical knowledge as part of an imperial project to enrich the Crown and improve Spain's colonies.[2] More immediately, they intended to reconcile their species identifications and organize their texts and images for publication as floras of Mexico and of Guatemala. They were determined to see the project to fruition, but fate ended up being almost as unkind to them as it had been to Hernández.

The archbishop of Mexico along with the cabildo (municipal council) of Oaxaca had donated funds to the Crown to support Sessé's organization of the findings into a *Flora mexicana* and *Flora de Guatemala*. But owing to economic and political difficulties in Spain—which was dealing with external wars, internal unrest, and disease outbreaks—the Crown never transferred these funds to Sessé, who ended up having to personally shoulder the cost of the project.[3]

It is unclear what became of Echeverría. In Spain, he remained close to the royal court, but suffered poverty and lamented his separation from his family in Mexico. Records in the Archivo de Indias show that in 1804 he received royal appointment as second director of painting at the Academia de San Carlos in Mexico City, but he likely was unable to return to New Spain to assume the post.[4] The last indication of Echeverría was in 1805, when he painted a prototype (see figure 21, item 0422 TCHI) for how the plants should be illustrated for the final version of the flora.

Mociño, meantime, became a rising star in European science, even as he suffered penury.[5] In 1805, he won a professorship along with induction into Spain's prestigious Royal Academy of Medicine in Madrid. With the death of Sessé on October 4, 1808, Mociño took over as head of the Royal Botanical Expedition to New Spain and continued his former mentor's task of organizing the findings for publication. Napoleon's invasion and occupation of Iberia that same year brought both opportunity and problems for Mociño. When the president of the Royal Academy of Medicine, José Martínez de San Martín, abandoned his post to join the resistance against Napoleon, and the vice president, Ignacio Ruiz de Luzuriaga, was arrested by the French armies, Mociño was elected as president of the Academy and as director of the Royal Museum of Natural History (Real Gabinete de Historia Natural). As president, he successfully petitioned for Alexander von Humboldt to become a member of the Academy, in recognition of the German's contributions to science during his travels in Mexico and the Andes. When the French withdrew from Spain in 1812–1813, the Spanish government ordered Mociño arrested as a traitor for having collaborated with the French during the occupation. Despite failing health and the gradual loss of his eyesight, Mociño packed a cart with the manuscripts and illustrations from the expedition and fled with the French troops to Montpellier, where his colleague Augustin Pyramus de Candolle offered him refuge.

De Candolle was so impressed by what he saw that, with Mociño's permission, he transported the expedition's natural history illustrations to Geneva for further study.[6] De Candolle explained that "I had only the pictures [*desseins*], for the text itself was lost on the flight from Spain." Despite the loss of most of the texts, De Candolle claimed to have "succeeded in describing and classifying" more than one thousand three hundred species "from the pictures alone."[7] From this work, De Candolle estimated that he identified around one hundred new genera and one thousand two hundred new species. In the process, he changed almost all the names that had been proposed by members of the expedition, selecting, instead names that honored men he admired, including eminent figures from Mexico such as Moctezuma and Archbishop Palafox, and key members of the Royal Botanical Expedition, including the draftsman Echeverría.[8]

When Mociño finally won permission to return to Spain in 1817, he asked De Candolle to return the material so that he could take it back to Madrid but agreed to reward his Swiss supporter with 305 of the illustrations that he considered duplicates. On the verge of losing access to most of the expedition's illustrations, yet eager to continue with his studies, De Candolle recruited over one hundred fifty Genevan artists to copy the expedition's images. He arranged his volunteers into teams, with the least experienced artists tracing outlines from the originals, experienced artists painting the colors, and the masters supervising. In fewer than eight days, they hurriedly copied 860 botanical illustrations and traced another 119. He dubbed the small army of mostly female artists the "Dames of Geneva."[9] As one might imagine, the varied levels of talent among the artists, the speed at which they labored, and their lack of access to plant specimens, resulted in botanical illustrations of uneven quality. To these De Candolle added the 305 duplicates that Mociño had given him, and another 71 that he had had copied back in Montpellier.

After Mociño retrieved his images from De Candolle, he headed to Madrid, but his health took a turn for the worse before he could complete the journey. He died in May 1820 at the age of sixty-three in Barcelona while under the medical care of Doctor Rafael Esteva, at the home of Jacobo Villaurrutia, whom he had befriended in Guatemala.[10] In cruel irony, the expedition's manuscripts and botanical illustrations became lost upon Mociño's death, meeting a fate similar to Hernández's images that the expedition initially had set out to recreate. For almost a century, the efforts of the Royal Botanical Expedition were known only via the work of De Condolle and the images produced by the Dames of Geneva.

Unlike the manuscripts and the illustrations, twenty thousand specimens from the expedition's herbarium survived in Spain. During this time, José Antonio Pavón Jiménez, a former leader of the Royal Botanical Expedition to Peru and Chile, stole many of the Mexican herbarium samples to sell for his personal profit.[11] According to Paloma Blanco Fernández Caleya, some of the samples stolen by Pavón simply became lost, while others ended up being acquired by the Botanical Conservatory in Geneva and the Botanical Institute of Florence. Some of these specimens in Geneva and Florence became type specimens to establish the nomenclature of the particular species they represented, but other specimens were impossible to identify because Pavón had removed their original tags. And the association of the samples with Pavón led some of them to be misattributed as native to Peru, rather than to Mexico.[12]

The expedition's herbarium, aside from the specimens pilfered by Pavón, was rediscovered in Madrid in 1934, deteriorated from neglect. The Spanish government sent it to the Field Museum of Chicago to be restored, catalogued, and studied by a team led by Paul C. Standley (1884–1963). After Standley and his team organized, mounted, and photographed the entire collection, they compared and correlated the specimens with the plant descriptions from the Royal Botanical Expedition's manuscripts, some of which had been rediscovered and published by the Sociedad Mexicana de Historia Natural at the end of the nineteenth century (see chapter 6).[13] The project was continued by Emilio Fernández Galiano (1885–1953) and eventually completed by Rogers McVaugh (1909–2009), who then transferred the collection back to Madrid in ninety-five boxes.[14]

In 1981, the Hunt Institute in Pittsburgh, part of Carnegie Mellon University, quietly placed a winning bid on two thousand botanical illustrations that had come to light a couple of years earlier in Barcelona. These turned out to be the long-lost original illustrations from Sessé and Mociño's Royal Botanical Expedition to New Spain. Apparently, upon Mociño's death in 1820, his doctor had taken possession of the images for reasons that remain unclear. He later sold them, and they ended up in the private collection of Luis and Jaime Torner. For a century and a half, the copies by De Candolle had been the most important tool for understanding the findings of the Royal Botanical Expedition. The reemergence of the originals created a stir, reigniting international interest in Mexican botany, a reshuffling of scientific names, and identification of many new plants. This, in turn, led to growing tensions among peasants, nationalists, and transnational pharmaceutical and agricultural companies over who had rights to these plants.

FROM ROUTINE TO UPHEAVAL: THE GARDEN AND SCHOOL OF BOTANY AND MEDICINE IN MEXICO

When Sessé departed for Spain, he brought along his wife María Guadalupe Morales, children Alexandro and Martina, and even his sister-in-law Josefa Morales. He anticipated that, after completing his task in Madrid, he and his family would return to Mexico, where he intended to serve as the permanent director of the garden, with Cervantes as the *catedrático* of the school of botany and medicine and López as the head gardener.[15] The commitment to Mexico on the part of Cervantes and Sessé owed something to their personal circumstances. Apart from his work in the garden and the school, Cervantes had accumulated wealth and prestige thanks to his successful private pharmaceutical business and his service as a physician at the San Andrés teaching hospital. His daughter and three sons were now pursuing their own careers in science and medicine within Mexico. Sessé, meantime, had married into a Mexican family and was benefitting economically from a lucrative local medical practice.

They had good reason for their optimism regarding their future in Mexico. When the celebrated German naturalist Alexander von Humboldt visited Mexico in 1803, just as Sessé prepared to depart for Europe, he was appalled at Mexico's social inequalities but stood in admiration of its scientific accomplishments. "No city of the new continent, without even excepting those of the United States," he wrote, "can display such great and solid scientific establishments as the capital of Mexico." He particularly praised Cervantes, Sessé, and Mociño, along with the expedition's botanical illustrator Atanasio Echeverría, whose illustrations "bear a comparison with the most perfect productions of the kind in Europe." Humboldt estimated that, thanks to the work of the expedition, the "new philosophies" of Linnaean botany and Lavoisierian chemistry were embraced more in Mexico than in much of Spain.[16] The Linnaean methodology had become firmly established in the colony, and the garden and school had entered into a productive routine. Cervantes regularly invited his Creole protégés back to the school to help teach the next generation of Mexican scientists and doctors. Together he and his former students cured the sick and expounded on the virtues of the many plants now being cultivated in the garden.[17]

The uprising that began with Miguel Hidalgo's September 1810 *Grito de Dolores*, along with Napoleon's 1807 invasion of the Iberian Peninsula, growing creole discontent with the Bourbon Reforms, and the development of a grassroots popular movement, set off a series of events that culminated in Mexican independence in 1821. More irksome to Cervantes than the general impact of the independence wars (1810–1821) and the occasional abuse that his workers and students suffered during their collecting expeditions was the callous treatment of the garden by Spanish troops. They commandeered the garden's space, dismantled many of its structures, destroyed the irrigation system and beds, and trampled the plants and soil that had been acquired with such difficulty.[18] Cervantes's dismay deepened when a wartime epidemic took the life of his long-time collaborator Jacinto López in 1813. The financially strained Spanish government refused to authorize a replacement, which Cervantes claimed led to the loss of hundreds more "plants that were vital to the lessons" in botany, medicine, and surgery.[19]

Cervantes's relationship with the Spanish government deteriorated further when a Spanish military official named Velasco erected barracks atop many of the garden beds, then tried to strip him and the garden of their functions. Velasco readily admitted that his troops had severely damaged the garden and destroyed its "choice plants." But rather than apologize, he argued that on account of this damage the garden could no longer even "provide the medicinal herbs most frequently necessary for the public." And so now that it could hardly even qualify as a "botanical garden" it was not worth its expense. Moreover, given the need to mobilize against the rebels, the government could no longer afford the luxury of scientific research. Velasco argued that the students who daily drained resources through their studies at the garden and its school would better serve the empire by joining the army and fighting in the front lines. He proposed that Cervantes be replaced by Diego Martín (dates unknown), who had been one of Jacinto López's assistant gardeners, at 16 percent of Cervantes's current salary.[20]

Outraged, Cervantes retorted that even Napoleon, during his aggression against Spain, saw fit to continue to fund science and learning, yet the Spanish Crown was proving unwilling to do so. He refuted Velasco's presumption that the garden was just some dispensary for medicinal herbs, declaring instead that it was a lighthouse of scientific learning and research. Though its collections had suffered at the hands of the Spanish troops, it still contained seven hundred species that, together with two hundred to three hundred plants brought in routinely from the countryside by assistants, enabled the sixteen to twenty remaining students to study medicine, pharmacy, and surgery while also conducting botanical and chemical research. Thus, it provided a benefit to society that he insisted was no less valuable than service on the battlefield. Cervantes particularly chafed at Velasco's suggestion that his role as head of the garden and catedrático of the school could be filled by Diego Martín, whom Cervantes dismissed as a poor, illiterate Indian who knew only enough to grow chiles, tomatoes, garlic, and onions on his chinampa. Cervantes persuaded the fiscal to reject Velasco's proposal but, in exchange, accepted further cuts to his already strained budget.[21]

By 1817, students and workers had resumed their collecting and research expeditions, and Cervantes even managed to coordinate an exchange of live plants with colleagues in Havana, Cuba.[22] The following year, he finally secured a successor for López. The new head gardener was the Italian Juan Lizardi (dates unknown, sometimes listed as Lazari or Lazary), who previously had overseen the private garden of Manuel Tolsá (1757–1816), the recently deceased head of the Academia de San Carlos.

By June 1821, two months before Mexico gained its independence from Spain, Cervantes already had ceased referring to the garden as the "Royal Botanical Garden," preferring to call it simply the "Botanical Garden." The documents do not reveal whether this was because Cervantes pragmatically responded to a shift in the political winds or whether it was because the royal forces had alienated him through their callous disregard for his scientific endeavors (or, more conjecturally, that his ill treatment by Spanish forces was the result of sympathies Cervantes may have had for the rebels, who included many of his former students). Whatever may have prompted the subtle name change, Cervantes seems to have tilted in favor of the republic.[23]

Though unable to fully recover from the losses caused by Spanish troops during the war, the garden was able to resume many of its activities in the years immediately following Mexican independence.[24] The new government, with its ambition to assert dominion over the territory previously defined as New Spain, apparently recognized the utility of the garden for claiming and ordering natural resources from across the country and placing these resources at the real and symbolic disposal of central authorities.

When the English naturalist William Bullock (c. 1773–1849) toured Mexico in 1822, he visited the garden in the viceregal palace and published this description in 1824:

> This beautiful establishment occupies one of the courts of the viceregal palace; and, though situated in the centre of a large and populous city, every vegetable production seems in perfect health and vigor. It affords to the stranger a most delightful retreat from the mid-day sun, and to the botanist, or admirer of the works of nature a treat not to be met with elsewhere in New Spain, or perhaps the world. It is handsomely laid out in the Spanish fashion, with flagged walks, bordered with elegant large pots of flowers. The

> walks are rendered cool by the creeping plants that are trained over them. They diverge from a large stone basin in the centre, constantly supplied by a fountain with water, which in small rivulets, spreads itself over every part of this little paradise—imparting freshness and life to thousands of elegant plants and flowers, unknown to the eye of an [*sic*] European, but which here, in a climate of eternal spring, in the open air, bloom and send forth their fragrance without the assistance of man, and produce a very different appearance to the dwarfish sickly exotics of our hot-houses, which, with every possible care and attention, with difficulty linger a few years without re-producing their species.[25]

In addition to revealing the balance between scientific and pleasure garden that the site had come to adopt, Bullock listed some of the plants and fruits that caught his attention, including "the most delicious sapotas [*sic*] I ever tasted. The celebrated hand-tree, which has excited so much attention among botanists, is in great perfection here. I have brought models of its highly curious fruit . . . as well as several species of the extraordinary cacti, mostly natives of Mexico, with which the garden abounds."[26]

Before leaving, he collected seeds from many of the plants, with particular interest in native flowering plants such as lobelia, ipomoea, dahlia, fuchsia, datura, and begonia, along with various kinds of cacti and euphorbias. "I hope in a short time to see them naturalized [in England], and flowering in our gardens and plantations." Despite Bullock's enthusiasm for what he saw, Mexico's financial problems led him to worry about the future of the garden.[27]

NATIONALIZING THE IMPERIAL PROJECT

As Bullock predicted, after his departure the garden suffered from the post-independence economic recession that hit most state institutions, until it found a champion in Lucas Alamán (1792–1853), the conservative Mexican politician, scientist, and former student of Cervantes who served as minister of foreign relations after independence. Alamán understood that post-colonial nation building required more than a political or economic grab at power; it demanded concrete visual demonstration of the natural-historical existence of the nation.

With this in mind, Alamán and Cervantes drafted a plan to reestablish the declining garden on firmer footing and combine it with a proposed museum of Mexican antiquities and natural history, aspiring to create an institution that would "rival or surpass establishments of [their] kind in Europe." After the fall of Agustín de Iturbide's brief Mexican Empire (May 1822–March 1823), the pair won state approval to move the garden to the cemetery of the Hospital de Indios (two blocks south of the modern-day Palacio de Bellas Artes, at the corner of Victoria and San Juan de Letrán; see map 3), where members of the Royal Botanical Expedition had conducted medical trials on indigenous patients. They began demolition of the chapels and walls and removal of the buried indigenous bodies to make space for native plants that the state now claimed as national patrimony. The government even started construction of an irrigation system, housing for workers, and paths for public viewing, as well as amendment of the soil.[28] Plans to move the garden's operations from the presidential palace to the hospital stalled after Alamán left the Ministry of Foreign Relations at the end of 1825. Nevertheless, initiatives continued to emanate from the cadre of scientists and politicians whom Cervantes had trained.

In 1827, the Mexican legislature endorsed anti-Spanish legislation that expelled all affiliates of the former royal institutions. But it carved out an exception for Cervantes, on account of "how grateful they were for his services and how deeply the loss would be if he were to leave."[29] At the time of his death two years later, on July 26, 1829, Cervantes was celebrated as a Mexican patriot.

By 1831, the combined botanical garden and its school of botany and medicine was under the charge of Cervantes's successor and former student, Miguel Bustamante (1790–1844). A congressional decree of that year ordered Bustamante and the heads of Mexico City's other scientific organizations to form themselves into a council headed by Pablo de la Llave (1773–1833) to promote scientific development and advise the government. De la Llave had worked as an assistant to Sessé during the Royal Botanical Expedition, and then moved to Spain where he continued his study of botany and theology, collaborated with Mociño, and even spent time in prison as punishment for his support of Mexican independence.

For the new generation of Mexican elites, which included many professional scientists and doctors, affiliation with the botanical garden became a sign of social standing and scientific learning. From 1821 through 1842, the head of the school of botany and medicine continued to deliver semi-annual lectures in the botanical garden of the Palacio Nacional.[30] Lizardi, "now 110 years old, bent double, but still in possession of all his faculties," was still working as head gardener in 1843, when the school of botany and medicine was converted into a department within the Colegio de Minería.[31]

After the death of Miguel Bustamante in 1844, the government appointed as director of the school his nephew, the botanist and zoologist Pío Bustamente y Rocha (possibly 1819–1886?)—whose father Benigno Bustamante y Septiem (1784–1858) had been a student of Cervantes, fought on the side of the royalists during the independence wars, and then was a founding member of the Sociedad Mexicana de Geografía y Estadística and a legislator. The council headed by Pablo de la Llave was dissolved in 1847, and the school now was placed under the Museum of Natural History and Antiquities, with Pío Bustamente as director of both the museum and the school until the school ceased operations around 1867.[32]

No doubt to Cervantes's postmortem chagrin, the survival of the gardens after midcentury in the presidential palace and in Chapultepec came to rest on their conversion into pleasure grounds for the rulers of the state and as symbols, rather than tools, of scientific modernity. Under the Hapsburg Emperor Maximilian who had been imposed on Mexico by conservatives during the French Intervention (1864–1867), the section of the garden on Chapultepec hill completed its transition into a pleasure garden, losing all traces of its previous scientific purpose, and it continued as such under the rule of Porfirio Díaz (president 1876–1880 and 1884–1911). The main garden in the presidential palace, meantime, took on the aura of historical curiosity. From 1878 through 1914, the Sociedad Mexicana de Historia Natural took over custodianship of the old botanical garden in the presidential palace, which remained in operation until the 1940s.[33] The garden never regained its scientific purpose or stature, but in the latter part of the nineteenth century a new generation of scientific intellectuals would revive many of its scientific objectives. Not content merely to pick up where the garden had left off, they would enlist state support to push the goals and methods in new directions, in the service of the modernizing Mexican Republic, as discussed in the following chapters.

CONCLUSION

By the mid-nineteenth century, the history of Mexican botany seemed like a procession of ambitious efforts that each ended in disappointment. During the era of Aztec rule, the native physicians known as *titicih* (sing. ticitl) had mastered the medicinal uses of flora from across their empire. We have no way to know to what degree they documented their knowledge, since the Spanish conquistadors incinerated almost all preconquest writings, some to purge Mexico of idolatry, others to intentionally cut off the indigenous population from its own history and sources of knowledge. Worried that indigenous botanical knowledge would be lost before it could be transformed into scientific and economic value, King Philip I sent his court physician Francisco Hernández to interview the remaining titicih, write down and test what they told him, and assemble a collection of simples and illustrations of New Spanish flora. Out of paranoia, the Crown locked away Hernández's manuscripts and the thousands of botanical illustrations produced by his indigenous assistants. In 1671, a fire in the Spanish royal palace destroyed the original illustrations along with the manuscripts. All that remained were distorted interpretations written by Nardo Antonio Recchi and Federico Cesi. Spain had squandered an opportunity.

Eager to reclaim fading imperial and scientific relevance during the Enlightenment, the Crown funded the Royal Botanical Expedition to New Spain (1787–1803) to update Hernández's findings and bring them into conformity with the teachings of Linnaeus. But this, too, met a sad end when the expedition's manuscripts and its trove of botanical illustrations went missing during the chaos unleashed by the Napoleonic wars.

Despite the loss of the material, the expedition and the institutions had an enduring impact on how scientific intellectuals and government officials within Mexico viewed their natural world. By renaming plants and removing them from their local cultural contexts, members of the Royal Botanical Expedition had sought to overwrite previous ways of knowing nature (especially the Náhuatl naming system and the expertise of the titicih), thereby rendering Mexico's natural resources "alien" to the very people who were most intimately familiar with them. Members of the expedition and their allies claimed the natural products of New Spain in the name of central authorities and asserted the exclusive right to rationally manage these resources in the name of the Spanish Empire, and then, after 1821, for the benefit of the Mexican Republic. They overtly devalued native ways of knowing nature and challenged local communities' claims to what post-colonial elites now insisted were *national* resources, even while they continued to rely on native knowledge and indigenous bodies. With all this, the Royal Botanical Expedition left a legacy and sets of claims that would be picked up and repurposed by nationalist scientific intellectuals during the last part of the nineteenth century, which is the third link in the chain and the subject of Part 3.

PART III

CLAIMING MEXICAN NATURE FOR THE REPUBLIC, 1868–1914

CHAPTER 6

THE NATURE OF THE REPUBLIC

In 1887, the prominent naturalist Gabriel Alcocer (1864–1916) declared "Mexico must recover what has been lost."[1] Among that which had been lost were the manuscripts and illustrations from both the Renaissance-era expedition by Francisco Hernández and the Enlightenment-era Royal Botanical Expedition, and, with them, the possibility to take advantage of those expeditions' discoveries. This was compounded by the many lost opportunities resulting from the political turmoil that had gripped the country since the time of its hard-fought independence from Spain, including devastating civil wars, an imperialist invasion in which the United States seized half of Mexico's territory (1846–1848), and a six-year period of French occupation (1861–1867). Starting in 1867, the republic at last entered a period of economic growth and relative political stability under the Liberal governments of Benito Juárez and then Porfirio Díaz. As the country's middle and upper classes gained confidence that Mexico could finally join the ranks of what they considered the world's most civilized nations, the country launched a forward-looking scientific project of nostalgic recovery. Past efforts that had served the interests of the Spanish Empire would be reanimated to claim nature as patrimony belonging to the modernizing Mexican Republic. And this time, Alcocer and others hoped, opportunity would not slip through their fingers.

The effort was spearheaded by scientific intellectuals affiliated with Mexico's two leading scientific societies: the Sociedad Mexicana de Historia Natural and the Sociedad Mexicana de Geografía y Estadística. Recent scholarship has added to our understanding of these societies' contributions to science, cartography, and geography, as well as to nation building. Scholars have produced insightful institutional histories of both organizations, as well as biographical and prosopographical (group biographical) studies of their members. This chapter and those that follow build on this scholarship but shift attention to environmental history. They address the ways scientific intellectuals from the 1860s through the 1910s engaged in forward-looking scientific nostalgia to channel and shape modes of seeing, describing, depicting, and exploiting nature (from individual plants to forests and even entire landscapes) in the service of the republic and in relation to indigeneity. These chapters also emphasize the crucial differences between these two organizations regarding how they each viewed Mexican nature and how they understood indigeneity.

During the mid-nineteenth century, foreign travelers had provided valuable descriptions and images of rural regions, but in the late 1860s Mexicans were becoming keen observers of their own countryside. For this reason, these chapters are concerned with individuals born and raised in Mexico, along with a small number of foreign-born scientists and artists who came to identify (and to be identified) with the republic and its landscape. In addition to analyzing how Mexico's scientific intellectuals viewed, organized, and depicted nature, and how they situated nature in relation to indigeneity, these chapters address how they tried to assert influence over government policies regarding the management of nature from the late-1860s emergence of triumphant liberalism through the early twentieth-century collapse of the Porfirian dictatorship.

THE EMERGENCE OF MEXICAN SCIENTIFIC SOCIETIES

In 1826, just five years after Mexico won its independence from Spain, the mathematician and naturalist José Espinosa de los Monteros (1776–1845) rallied politicians and intellectuals to create the Instituto Nacional de Ciencias, Literatura y Artes.[2] A few years later, in 1831, the government chartered the Museo Nacional Mexicano (originally spelled "Mejicano," with a *j* instead of an *x*), which incorporated both antiquities and natural history, including the remnants of the old Royal Botanical Garden.[3] Despite these tepid early efforts, it was not until 1868, prompted by Benito Juárez's triumph over the French Intervention, that Mexicans took two decisive actions in support of science and learning.

The first of these two 1868 advances was the creation of the Sociedad Mexicana de Historia Natural (SMHN), dedicated to the study of Mexican nature. Membership required education or accomplishment in the field of natural history and skewed especially toward individuals with medical training. It even counted the eminent landscape painter José María Velasco among its leaders (for reasons that will become clear in the coming chapters). To publicize its scientific findings, the SMHN launched a journal called *La Naturaleza* in 1869, which happened to be the same year that the identically named scientific journal *Nature* was launched in London.[4] The SMHN was an independent scientific society supported by membership dues, though it also benefitted from state subsidies and commissions.

The second 1868 advance was the reorganization of the Sociedad Mexicana de Geografía y Estadística (SMGE). The SMGE was older than the SMHN, having been chartered in 1833 with the broad mandate of promoting geographic and statistical knowledge that might contribute to the prosperity of Mexico. It was the first geographic society in the Americas and the fourth in the world.[5] Unlike the SMHN, the SMGE was fully dependent on the government, operating as part of the Secretaría de Fomento (Ministry of Development).[6] It published a professional journal titled the *Boletín de la Sociedad Mexicana de Geografía y Estadística* and, like the SMHN, counted among its members some of the leading scientific intellectuals of the day.

By restructuring the SMGE in 1868, members sought to distance their organization from its earlier collaboration with the French imperialist government. To emphasize its fresh start and its new commitment to the Mexican Republic, the SMGE relaunched its *Boletín* in what it called its "*segunda época* (second era)." In the coming decades the *Boletín* published on a dizzying range of subjects from astronomy to zoology, with particular interest in the standardization of measurements, collection of statistics and scientific information from across the country, creation of a national map of Mexico, and the stimulation of large-scale export-oriented commercial agriculture.[7]

Both the SMHN and the SMGE used their professional journals—which they sold cheaply within Mexico and distributed to partner organizations around the world—to expand their influence. Unlike the SMGE's *Boletín*, which had a very broad mission and whose articles were written by a mix of politicians, scholars, and dilettantes, the SMHN's journal *La Naturaleza* focused specifically on natural history, with special attention to things native to Mexico, and it required that all its contributors possess scientific or medical training.

In its mission statement, the SMHN's *La Naturaleza* described itself as a "journal concerned almost exclusively with our own land, so rich in natural products of every kind, many still unknown by the scientific world." It stressed that instead of just serving the elite and the national treasury it was dedicated to discovering "which of these [native natural resources], properly exploited," can "improve the fortunes of our patria by providing work and well-being to the innumerable people that currently lack life's most basic resources." It invited all "enlightened people" from across the country who are "lovers of their patria," including members of the government, to support the SMHN in the pursuit of its agenda.[8]

From 1869 through 1914, *La Naturaleza* published research into native flora, native animals, and minerals. Among these, flora held particular importance, anchoring the nation to its indigenous nature and presenting the possibility of cultural and economic autonomy. The SMHN's attention to native species was rooted in neither xenophobia nor an obsession with purity but rather sprung from a nationalist attempt to better utilize underappreciated native species as sources for Mexican pride and economic development. Its publications and activities also point the way toward an inchoate appreciation of local ecosystems.

While members of the SMHN and the SMGE embraced one another as colleagues, and even had some minor overlaps in membership, the following chapters reveal the importance of differences between the two organizations. These differences led the SMHN and the SMGE down sometimes-conflicting paths, with enduring implications for how nature and indigeneity were understood within Mexico.

APPROPRIATING LOCALITIES FOR THE REPUBLIC

In the post-1868 era, the SMGE was eager to redeem itself after it had collaborated with the government of Emperor Maximilian during the French

Intervention. Continued anxiety over Mexico's loss of half its land during the Mexican-American War (1846–1848), combined with the ongoing dispute with Guatemala regarding the Soconusco region (see map 2), inspired the reorganized SMGE to demonstrate that it could act in defense the republic's territorial integrity. It did so by sending expeditions to collect information from across the country, including some of its most distant corners. The work of these expeditions was supplemented by regional councils across the country who collected and shared local information in support of the SMGE's goals, including descriptions of local natural environments that might be exploited for the economic growth of the republic.[9]

The geographical descriptions published in the *Boletín* served both scientific and ideological purposes. SMGE members explicitly stated their hopes that the information would also serve economic ends by identifying locally available resources such as vegetable-based wax in Cautlali or coffee and cotton in Xalapa that could be profitably exploited, and which might even lure foreign investors.[10] In these efforts, SMGE members demonstrated indifference to whether plants were native to the region or even to Mexico, but they did take an interest in the ancient indigenous heritage of each region.

Race and history became woven into their discussions of localities. This was the case whether authors were addressing what resources each region offered, who had the right to control those resources, or what to call a place or a natural feature. As historian Raymond Craib and others have pointed out, Mexican leaders of the first half of the nineteenth century had demonstrated a general "indifference to Mexico's indigenous heritage."[11] Historian Erika Pani has similarly noted that:

> Until the 1870s, the indigenous element did not appear in academic paintings nor in architectural motifs. It rarely figured in poems, short stories, or novels that claimed to contribute to a national culture. Writers preferred, as mockingly confirmed by Vicente Riva Palacio "for the plots of their stories, characters from the [European] Middle Ages who love and fight in the fantastic castles along the Rhine . . . without ever daring to create space in their tales for cuitlacoche [huitlacoche, a corn mushroom common in Mexican indigenous cuisine], the cenzontle [a native songbird], cacomite [a native species of lily], nor yoloxóchitl [a native magnolia]."

And, "outside of a small circle of experts, pre-Hispanic ruins" were treated merely as "mounds of stone that could be used for construction or sold."[12]

Starting at the end of the 1860s, the SMGE began to eagerly recover this neglected ancient indigenous heritage. It did so in a manner that validated ancient indigenous cultures while denigrating contemporary indigenous people, thus establishing non-Indians and the Mexican Republic as the rightful heirs to preconquest civilizations, to the land and resources those preconquest civilizations had once controlled, and to the knowledge these native peoples had produced about their natural world.

Two regional reports published shortly after the SMGE's 1868 reorganization provide examples of how indigeneity was being inscribed on the landscape in a manner that separated the land, natural resources, and ancient Indian civilizations from the indigenous people who still inhabited the space. The first example is a report on Sonora and Baja California written by Manuel Payno (1810–1894, military veteran, historian, writer, teacher, politician, and naturalist), which started by telling readers that the "civilized population" of this region consisted only of mestizos and some recent arrivals from Spain. As for the resident Opate Indians, Payno rejected their claims to be indigenous on the basis of his own difficulty in distinguishing them from their mestizo neighbors. His rejection of their indigenous status did not lead him to accept them as part of the "civilized population."[13] Instead, he placed them outside the tide of history, orphaned both by their own past and by the future-oriented Mexican Republic. The second is the 1869 publication of Federico Aldherre's report on Yucatán, which even cast doubt on whether Mayans were really Indians. Aldherre (dates unknown), who had accompanied Empress Carlotta (1840–1927, empress consort to Maximilian from 1864 to 1867) on her trip to the peninsula in 1865, claimed that any real Indians in the Yucatán had already "disappeared" long ago. He dismissed any claims to the contrary by locals as mere cover for opportunism, banditry, and unjustified racial hatred against the white population. Despite rejecting Mayans as genuine Indians, Aldherre referred to them as "*indios*" in disparaging ways throughout his essay, claiming, for example, that "these miserable Indians know nothing and confuse and distort everything" as a result of their "barbarism" and "ignorance."[14] Thus, they were at once Indian and not-Indian: they possessed all the negative traits that Aldherre ascribed to Indians, but any grievances or claims to land or natural resources on the basis of their heritage or as a result of racism against them were discounted on the grounds that they were only pretending to be Indians. Other reports, meanwhile, made clear that indigenous people who overtly resisted Mexican authority were entirely denied any claims on the republic or its territory, and even pressed into slavery on henequen plantations. This included groups such as the Apache of Sonora, who reportedly lived in the mountains and came into Mexican territory from "their lands" only to rob and pillage Mexican citizens.[15]

SMGE member Carlos de Gagern (1826–1885) went even further, cautioning his fellow citizens that being a true patriot meant not closing your eyes to the fact that indigenous people were "retarding the development of Mexico," and that they therefore must be removed as obstacles.[16] The real tragedy, he argued, is not the elimination of the "inferior races," but when a "vigorous youth is chained to a decrepit old man," or when a Pegasus finds itself yoked to an ox. Is it any wonder, he asked, that the Pegasus does not soar, but instead, yoked to the ox, it goes round and round with no progress? This, he argued, was the situation in which Mexico found itself, with indigenous people as the "main cause of Mexico's misfortunes." Frontal efforts to destroy indigenous peoples' identities would fail. Instead, "it is necessary to find levers to lift this inert mass and put it in motion, and not leave the Indian with time to sit with his arms crossed" watching "the current of civilization flow past him." Mexico must force him into the current, so that it either drags him to his own improvement or drowns him. Gagern proposed that a major obstacle to his plan was Indians' deep racial hatred toward white people, who they "unjustly" blamed for their loss of land. According to him, the current-day Indians had no right to make any claims on the land because of their own backwardness, lack of ambition, and disinterest in progress that rendered them incapable of exploiting that land.[17]

Narratives of indigenous decline made their way into most studies of native people at the time. It even shaped the study of indigenous languages. Eufemio Mendoza (1840–1876, lawyer and linguist), for example, urged his fellow intellectuals to take an interest in Mexico's indigenous languages, but not the way they were spoken by contemporary native people, since the once-noble languages had been corrupted by modern-day Indians. After cleansing the languages, intellectuals could teach the indigenous communities to properly speak their own languages, and then even use them in publications, so that "the Indian, today wretched and isolated, might enter fully into civilized life, cease to be a foreigner in his own homeland, and join with a million other true citizens." Spoken to in a rectified version of their own tongue, "within short order, [Indians] will become real citizens, and this will make Mexico great." Even the Yucatecan bishop, ethnologist, and linguist, Crecenscio Carrillo (1837–1897), who insisted that Mexico's preconquest civilizations had possessed a vibrant literature, shared the assumption that such a thing no longer existed owing to the backwardness of modern-day Indians.[18]

Validation of ancient indigenous civilization, paired with denigration of modern-day Indians, extended into the study of botanical wealth, architectural sites, and the landscape itself. Whites in the United States imagined themselves as the inheritors of a mostly empty wilderness, but in Mexico land was heavily marked by history and filled with indigenous people who, by virtue of their participation in the wars against Spain, France, and the United States, and armed with their *titulos primordiales* (original land titles that had been issued by the Spanish Crown), could assert their own claims to the land. Scientific elites of the SMGE addressed this challenge by developing a tailored discourse for expropriation.

We can see how this operated by looking at an article that Eduardo Ruiz (1839–1902, military veteran, politician, and amateur ethnologist) published in the *Boletín*, in which he insisted that the archaeological sites in Michoacán, of "whose foundation and history the [modern-day] Indians are ignorant," prove that ancient peoples of the region had achieved greatness, and that modern-day Michoacanos could do so once again. The rightful owners of the land and its natural resources were the mysterious ancients—whose mysteriousness and separation from current indigenous people are essential for his argument. In their absence, this inheritance passed to whites and mestizos who could harness the land on a level equal to or superior to those ancients, rather than to modern-day indigenous Tarascans (Purépecha), who he argued were incapable of exercising the responsibilities of modern citizenship and industry. In this way, Ruiz discursively wrested the land and its natural resources from the indigenous population, placing it in the hands of men such as himself, whom he saw as the ancients' true heirs.[19]

SMGE reports engulfed places in carefully curated narratives about history, geography, race, and power, to claim these places on behalf of the modernizing republic. Detailed descriptions of a location and its history were not enough. Names themselves had to be changed, returning them to their ancient indigenous origins. Enlightenment botanists had used Linnaean binomials to *free* plants from their historical, cultural, and territorial contexts so that they could be claimed by the distant metropole.[20] SMGE members, by contrast, sought to use names as a way to anchor places, natural features, and native plants to ancient indigenous histories, so that they could be claimed as part of the national patrimony of the Mexican Republic, the supposed heir to those ancient civilizations.

This required recovery of the "original" indigenous names of locations, natural features, and, increasingly, plants.[21] Examples of Náhuatl names (along with their meanings) preferred by the SMGE over Spanish-language alternatives included Cuauhtitlán (next to the eagle), Chicomoztoc (seven grottos), Apam (over the water), Chicontepec (seven hills), and Chapultepec (grasshopper hill). Tarascan names from Michoacán included Acámbaro (place of maguey) and Copándaro (place where avocados abound).[22] These indigenous place names favored by the SMGE referred to each locality's natural environment and plants rather than to human-centered histories or practices. While scientific intellectuals at the time could have found any number of ways to stabilize place names for recordkeeping, historical interpretation, and land titles, it was no accident that their solution was to immortalize autochthonous ties between place and nature, thus affirming the republic as the inheritor of this deep history and primordial landscape.

Debates of the time not only addressed what name should be used for a place, but also how to spell and pronounce that name. In particular, they debated the question of whether to use *x* or *j* in place names such as Xalapa and México. In 1870, Manuel Payno forcefully argued in favor of the *x*. He insisted that if Náhuatl had been common in a region during any period of its past, because of the expansion the Aztec empire, then the place-names within that region should take the *x* instead of either *g* or *j*. By using the *x* to emphasize that a place had once been part of the Aztec empire, Payno affirmed the right of the republic, as rightful heir to the Mexica, to assert its own authority over the locality.[23]

Alongside the effort to study and dominate the diverse localities claimed by the republic, there emerged a growing interest in whether particular plants were native to Mexico, and how these native plants might fit into Mexico's past, present, and future.

"MEXICANS GAVE NAMES TO THINGS": BOTANICAL PATRIOTISM AND NATIVE PLANTS

The SMGE's assertions of dominion over indigenous land were soon followed by efforts by members of the SMHN to lay claim to Mexico's plants, but with a more positive view of native people and their cultures. In 1876, the physician and linguist Gumesindo Mendoza (c. 1829–1896) and the pharmacist Alfonso Herrera (1838–1901), both of whom were members of the SMHN, published their botanical analysis, or monograph, of a tree called anacahuite. Their text is an instructive example of how scientists were beginning to elevate the sixteenth-century Hernández expedition as a point of linkage between preconquest knowledge and modern patriotic science. Rather than treat the plant as an isolated specimen liberated from place and culture, as often was done by eighteenth-century followers of the botanist Carl Linnaeus, Mendoza and Herrera explicitly endowed the anacahuite plant with historical and cultural context.

They started by informing their readers that, thanks to Hernández, we know that "anacahuite" is derived from the Náhuatl name *amacuahitl,* which means "paper tree": "From its etymology we can reasonably infer that the ancient Mexicans gave names to things, indicating with them one or some of their properties or uses." That "ancient Mexicans" figured out these properties and uses through careful observation and experience demonstrates that "our ancestors were dedicated to the study of nature. We have a multitude of evidence with which to corroborate our assertion."[24] With this, Mendoza and Herrera validated ancient indigenous people as capable of science, while, via Francisco Hernández, appropriating them and their knowledge as "ours."

They refer to Hernández in a telling manner:

> Hernández, the wise doctor of Philip II, came at the end of the 16th century to explore these vast and rich regions; for seven years he walked with admirable perseverance, acquiring from the mouths of indigenous doctors what they knew about the properties and uses of plants, and he collected,

> so to speak, the last glimpses of a civilization, which was disappearing amid insatiable thirst for gold and before the religious fanaticism that demolished the temples and statues covered in hieroglyphs, and threw into the flames the manuscripts that contained the history of a great people and the secrets of the sciences.[25]

Hernández thus became a model for how to value and build on Mexico's indigenous cultures and native flora.

Mendoza and Herrera's narrative is noteworthy for two reasons. First, in contrast to Enlightenment-era scientists who had sought to strip plants of accidents of place, culture, and history, Mendoza and Herrera were part of a trend among Mexican scientists in which they recovered historical narratives and indigenous provenance as fundamental to their modern botanical analysis. They made this kind of historical and cultural context, as well as reference to indigenous people and to Hernández, a standard part of botanical analysis within Mexico. Second, Mendoza and Herrera's celebration of Hernández's efforts to preserve indigenous knowledge, along with their condemnation of Spanish colonists, was part of a shift in how Mexicans viewed their history toward a more charitable view of indigenous culture.

As Mendoza and Herrera illustrate, it was not enough for Mexico's natural scientists to position themselves as the inheritors of European and Spanish imperial Enlightenment science: they also had to show themselves as heirs to and defenders of Mexico's indigenous intellectual traditions and knowledge, an essential part of what distinguished Mexico from Europe. Archaeologists and museum professionals had made significant progress in the study and appropriation of ancient indigenous artifacts, architecture, and history as part of Mexico's nationalist patrimony.[26] But the question of how to root the nation in indigeneity was a more complicated question for natural science than it was for archaeology, a question that naturalists of the time attempted to answer.

Notably, though Mendoza was indigenous and his first language was Otomí, he and Herrera were reluctant to extend their praise for ancient Mexicans to modern-day indigenous people. Along with most of their colleagues and in partial agreement even with SMGE member Francisco Pimentel, they saw modern indigenous people as backward.[27] But whereas Pimentel insisted that indigenous culture had little to no redeeming value beyond its symbolism, Mendoza and Herrera were part of a growing shift led by the SMHN in which they argued that indigenous people had been victims of an intentional campaign of destruction by rapacious Spaniards, but that they and their ancestors possessed valuable scientific knowledge, some of which persisted into the current era. The lesson for scientists of their own day was clear: emulate Hernández by documenting indigenous botanical and medical knowledge, then build on this knowledge for the good of the nation and for the broader world.

Botanists within the SMHN became convinced by Mendoza's and Herrera's view regarding indigenous medical knowledge, but many members of the SMGE continued to insist on the need to denigrate contemporary Indians. In his analysis of a native bindweed, Carlos Casas (dates unknown) of the SMGE, for instance, argued that Mexico's current indigenous people were "degenerated and crude" and had lost all connection to the great ancient civilizations that had once cultivated botanical and medical knowledge. And so it was up to modern men of science such as himself to recover this lost knowledge for the economic benefit of the Mexican Republic and its regions.[28] Casas's SMGE colleague Eduardo Ruiz was only slightly more sanguine about contemporary indigenous people when he argued that a minority of Tarascans in Michoacán still preserved valuable botanical and medical knowledge inherited from their forebearers. The majority, however, chose ignorance over science by embracing witchdoctors rather than utilizing the traces of legitimate knowledge handed down from their ancestors. As scientists identified and extracted this Tarascan knowledge, he argued, they were discovering that Michoacán's botanical riches, ranging from fruits and crops to lumber and medicinal plants, rivaled its abundant mineral resources.[29] And so, in a vision of himself as a modern-day Hernández that was more contorted than anything propounded by members of the SMHN, Ruiz took up the responsibility of extracting what little botanical knowledge remained among the Indians about native plants before it was lost forever.

When it came to agricultural matters, members of the Sociedad Mexicana de Geografía y Estadística, unlike those of the Sociedad Mexicana de Historia Natural, usually expressed disinterest in whether particular plants were native, and even showed a preference for foreign crops. In his 1871 article in the SMGE's *Boletín*, the *jefe político* (the local strong-man imposed by the central government) of the Soconusco region of Chiapas, Sebastián

Escobar (1831–1893), listed plants of potential economic value within that zone, a territory that Mexico and Guatemala each claimed as its own. But Escobar showed no interest in distinguishing native from nonnative plants. Even for cotton and cacao, which he described as the region's most valuable crops, Escobar made no mention of the fact that both were native to the region.[30] Pedro Mendiondo went further by invoking a nationalist spirit specifically in support of nonnative crops. He celebrated the efforts by the SMGE to promote imported crops as acts of "true patriotism" because these nonnative crops could help free Mexico from dependence on imports, and perhaps one day even elevate Mexico "to the rank it deserves."[31]

If coffee from Arabia, mangos from China and Japan, and quinine from Peru can be acclimated to our land, Mendiondo wondered, why not tea, too? And why stop there? He listed other agricultural products that could take similar advantage of the many climates found in Mexico. And if plants could be acclimated for the economic benefit of the country, then so could European immigrants, he argued. This recalls the situation in Argentina or Australia, where historians Antonio Brailovsky, Dina Foguelman, and Thomas Dunlap have found that acclimatization of foreign plants and people were central components to the process of domination, as leaders intentionally overwhelmed native species and native peoples with plants, animals, and white people imported from Europe.[32] Similar to their Argentinian and Australian counterparts, members of the SMGE hoped to solve Mexico's economic woes by transforming the country with imported crops and European migrants.

The SMGE's dream of overwhelming native plants with nonnative commercial crops met resistance from the SMHN, whose members argued that native plants could be as profitable as imported crops and were more appropriate to the Mexican landscape. The SMHN geologist, historian, and botanist Mariano Bárcena (1842–1899), for example, shared Mendiondo's enthusiasm for acclimating plants, but rather than import exotic species he preferred to move plants from one part of Mexico to another, and showed a more subtle understanding of what a plant species might need to thrive. To figure out which Mexican plants could be productively acclimated from one region to another, Bárcena encouraged botanists to pay closer attention to isotherms, soil chemistry, hydrology, and what today we would call microclimates.[33] He urged his colleagues to consider not just whether a plant could grow in a particular region, but also what kind of effect the plant would have on the local environment. In this inchoate manner, Bárcena pointed the way toward the need to study the ways invasive plants might negatively impact the local ecology.

The editors of *La Naturaleza* paired Bárcena's article with a contribution by the French naturalist Carlos Martins, which drew on recent findings about ice ages to analyze how changes in the land and climate had contributed to the natural migration of plant species within Mexico.[34] This appears to be one of the earliest cases of historicizing the Mexican landscape by asking how the geographical distribution of botanical species varied over time, which was a departure from the static view of nature clung to by creationist-inclined scientists of the era. Read together, the articles by Bárcena and Martins suggest that acclimatization of plants from one part of Mexico to another might be understood as an extension of the natural migration of species and consistent with the protection of the organic integrity of the Mexican republic.

Scientists were not just concerned with how to move plants or alter environments. They also were eager to understand how Mexicans already were exploiting native plants in their natural habitats. It was with this goal that Manuel Payno, who was a member of both the SMGE and the SMHN and a bridge between the two organizations, penned a study of maguey, which already was emerging as an iconic botanical symbol of Mexico. He called on botanists to classify the various kinds of agave (including maguey), chemists to study the parts of the plant and their uses, and medical researchers to figure out pharmaceutical uses of maguey and its products. He also recounted nationalist stories and myths connected to maguey, thus rooting the plant firmly within Mexican soil, history, and society.[35] Vicente Fernández drew inspiration from Payno in his similar study of mezcal for *La Naturaleza*.[36] Others followed by studying a range of additional native plants already being commercially exploited.

To advocate for native species, scientists had to figure out what it meant to be native, and which plants qualified. Sometimes the question was about whether a plant was originally from a specific region. In his 1872 study of how to increase cacao production in the southeastern state of Tabasco, the local politician and amateur naturalist León Alejo Torres (1831–1895) agreed that cacao was native to Mesoamerica, but speculated that the lack of a distinctive local indigenous name for it left it uncertain whether the tree was native to the state of Tabasco.[37] It is also notable that one of the important clues he was searching for was the original indigenous name. We see this again in an article by V. Reyes (dates unknown) published later that year, arguing that the legume tree *tepehuaje* was a valuable but underexploited source of lumber. Considering the name, paired with earlier observations by Alexander von Humboldt, Reyes concluded that not only was the tepehuaje tree native to Mexico, but more specifically it originated in the Pacific coastal region around Acapulco. He urged that if producers could harvest tepehuaje trees at fifty years rather than the usual four years, they could produce high-quality lumber, thus enabling Acapulco to exploit this native plant to compete domestically and internationally with the best-quality lumber imported from Europe.[38]

The historian Alfred Crosby has shown how rapidly the exchange of plants, animals, and diseases proceeded after initial contact between the Old World and the Americas.[39] The early and rapid occurrence of such exchanges sometimes left Mexicans of the late nineteenth century scratching their heads about whether something was native to Mexico. The physician, pharmacist, and naturalist Alfonso Herrera attempted to address one of these puzzles in 1882 when he wrote about the banana. Despite poring through historical documents, he was unable to definitively resolve whether the banana was native to Mexico, or a very early introduction from a distant corner of the Spanish Empire such as the Philippines.[40]

Debates consistently operated under the presumption that the outline of the Mexican Republic during the late 1800s, not the reach of New Spain, much less the broader Spanish Empire, marked the boundaries of what counted as native. Botanists also ignored Mexico's pre-Mexican-American War boundaries, which had included areas taken by the United States in 1848. Rather than acting as a source of pride, those old national boundaries perhaps invited painful memories of Mexico's humiliation at the hands of its northern neighbor.

Imperial claims by the Aztecs, centered in Mexico City, were a different matter. In 1871, the prominent intellectual Ignacio Ramírez (1818–1879), who in other contexts sometimes went by the nom de plume "El Nigromante," wrote admiringly in the SMGE's *Boletín* about how the Aztecs had concentrated in their capital cacao, gold, and other tribute from across their empire, along with "exquisite plants for" display in "the gardens of the kings."[41] Mexico City/Tenochtitlán served as capital of the Aztec empire and of the Mexican Republic, in both cases acting as a centrifugal hub from which to assert authority over the surrounding regions, as well as the centripetal hub that attracted resources and wealth toward the imperial-cum-national center.

According to late nineteenth-century nationalists, the garden created by the Royal Botanical Expedition (see chapter 3) had inherited the role of the Aztec gardens by collecting plants from across the empire for display in Mexico City. But to make the former imperial garden serve the patriotic needs of the republic, scientists had to engage in two kinds of regeneration. First, they had to define the Royal Botanical Garden as a center for early nation building rather than a tool of Spanish colonialism. Second, they had to deal with the remains of the garden, which had become a faded image of its former self (see chapter 5). The time seemed right to replace it with a new garden that could better serve the changing needs of the Mexican Republic.

INVENTORYING MEXICAN NATURE

In his 1873 speech commemorating the five-year anniversary of the Sociedad Mexicana de Historia Natural, the organization's *secretario* José Joaquín Arriaga announced the creation of two special committees. The first would advocate for the creation of a new botanical garden in the capital in compliance with Article 7 of the Ley de Instrucción Pública of May 15, 1869. Such a garden, according to Arriaga, would showcase for scientists and the public "the great beauty of the plant kingdom from the diverse regions of the Republic." The second committee would promote the study of Mexico's "botanical geography." This would involve compiling a comprehensive list of all native plants, information about species (both native and foreign) being agriculturally exploited within the country, and climatological data to determine where particular plants could thrive.[42] These two efforts to inventory and organize Mexican nature—a national botanical garden and a botanical geography—would focus much of the SMHN's energy during the coming decades.

Fourteen years later, in 1887, Gabriel Alcocer expressed frustration about the lack of progress on the garden. Alcocer was sympathetic to the fact that political difficulties of 1869 had prevented the government at that time from complying with this as well as other parts of the law—such as creation of an astronomical observatory and an Academia de Ciencias y Literatura, along with expansion of public education at the primary, secondary, and professional levels—but he felt that the stability of recent years no longer left any excuses.[43] Foot-dragging on this project, he claimed, slowed economic growth and hindered the effort to attract a new generation of students to the science of botany. It even threatened national autonomy because Mexicans now knew less about their native plants than did foreign interlopers.[44]

Attempting to shame the government into action, Alcocer asserted that "it is necessary and sad to confess that Mexico owes more to Spanish domination" than to its own national government. The Spanish Empire, he claimed, had acted promptly regarding the Royal Order of November 1787 that called for the creation of the Royal Botanical Garden and school of botany and medicine in Mexico City, of which the Mexican Republic ended up being a beneficiary. The current government, by contrast, continued to drag its heels in relation to the Ley de Instrucción Pública of 1869.[45]

Alcocer proposed Chapultepec as the ideal location for such a garden, unwittingly reiterating each of the arguments that Sessé had made for the site a century earlier. He also suggested that Mexicans of all walks of life could contribute to its collection. "Indians who dedicate themselves to this kind of trade" could bring native plants "that are well known to them" from all across the republic. Garden employees would seek out plants at native markets around the city and in other parts of the country. State governors and regional officials could send seeds and live plants as demonstrations of their patriotism. And scientists, backed by the power and finances of the state, would oversee the entire process, while assuring that the garden contributed to both national and global scientific knowledge. Once it was all assembled, the general public could come witness the beauty of Mexican nature while learning about modern science. All this, taken together, would assure that the garden would serve as a truly "national project."[46]

The second initiative that José Joaquín Arriaga had announced back in 1873 was for a botanical geography of Mexico. This was a more multifaceted initiative than the botanical garden, yet easier to make progress on. The SMHN immediately began publishing in *La Naturaleza* a list of all known native plants. It ended up taking a decade (1873–1883) to publish the entire list, one section per issue, with the understanding that research expeditions would continue to bring to light more plants that would need to be added.[47]

Tellingly, the plants were not alphabetized by their Linnaean Latinate binomials, but according to their vernacular names, most of which were derived from indigenous languages. Each plant's common name was followed by its Linnaean binomial to show how it also fit within universal science; but it was the Mexican common name that received priority and was the basis for organization. In this way, the authors affirmed that what plants were called mattered.

In 1876, the SMHN began to put the inventory to use when it represented Mexico at the Philadelphia world's fair with an exhibition of native trees and plants sorted by economic use: lumber, oils, resins, dyes, and fibers for textiles. Later that year, *La Naturaleza* published the list of plants that Mexico had contributed to the Philadelphia exhibition, along with two additional lists of native medicinals (*drogas indígenas*) that included the common and Latinate name for each plant, where it grew, what it was used for, and, making explicit the economic aims, the current price per kilogram.[48] After the SMHN's exhibition in Philadelphia won an international award, editors of *La Naturaleza* were inspired to publish additional plant inventories, following other systems of organization such as type (trees, flowers, legumes, etc.), economic uses, and region of origin. These plant lists, even in their incompleteness, became aesthetic and nationalist objects in their own right, and testaments to the natural integrity and indigeneity, as well as the cosmopolitanism, of the modernizing Mexican Republic.[49]

Members of the SMHN who were trying to assemble a comprehensive national flora, or plant inventory, pored through European archives in search of the lost manuscripts from the Royal Botanical Expedition.[50] They finally located one of these texts in 1871 and over the coming decades found more, bringing each to publication for the first time, with implications that will be discussed in the chapter's next section.

DOCUMENTING MEXICO'S SCIENTIFIC LINEAGE

While nineteenth-century scientific intellectuals praised the Royal Botanical Expedition for placing Mexico into the current of modern science, it was the older undertaking by Francisco Hernández that they felt anchored the republic to its indigenous nature. The physician José Eleuterio

González Mendoza (1813–1888) of the SMHN praised Hernández for having saved "from oblivion not just the indigenous names of animals, minerals, and plants of the region, but also Aztec medical traditions, since in describing and naming each thing," Hernández also "indicates the uses the Indians made of it." He argued that, after Hernández, Spanish natural history fell into decline for the next two centuries, until it was revived by Sessé and Cervantes and their Mexican "disciples," who included "Mociño, Maldonado, Bustamante, Cervantes (son), Larreategui, Bernat, Peña and Monroy." González situated the SMHN as the inheritor of this tradition, arguing that "Hernández lay the foundations for Mexican Natural History, Cervantes and his many disciples cultivated it with diligence; and the Sociedad Mexicana de Historia Natural brought it to the level found in the most learned nations of Europe."[51]

To make clear the importance of this legacy, some of the SMHN's botanical descriptions (monographs) even elevated the preliminary Latinate binomials proposed by Cervantes, Sessé, or Mociño over the binomials that European scientists ultimately settled on. By doing so, they emphasized the importance of the science done within Mexico rather than simply defer to European authority. In 1873, José Morales (1850–1929, medical chemist and professor of toxicology), for example, in his analysis of the plant that he tells us is known in Náhuatl as *ypacihuitl* and in Spanish as *yerba de zorillo*, made no mention of the Linnaean binomial resolved in Europe (i.e., *Petiveria alliacea* L.). Instead, he offered the Latinate names proposed by members of the Royal Botanical Expedition: *Croton Vulpinum* and *Croton Dioicus*. Using modern chemical analysis of the plant, Morales confirmed that it exhibited the medicinal properties described by Indians. In this way, he rooted the plant in Mexican culture and soil with a lineage that ran from preconquest civilization through Hernández and the Royal Botanical Expedition. And he affirmed the plant as national patrimony linked to modern-day indigeneity and claimed it as a resource that could be understood and exploited specifically by Mexican scientists for the economic benefit of their republic.[52]

With each plant monograph Mexican botanists recovered more of what had been lost, reaching back to Hernández with a consistency reminiscent of the way Renaissance botanists had regularly invoked Pliny or Dioscorides. Alfonso Herrera's 1873 analysis of a type of oleander is a typical example of the SMHN's distinctive approach to describing and claiming native species. Herrera's plant monograph starts with a romantic description of the beauty of the plant situated within the fertile mountains of Mexico. It firmly roots the plant within indigenous culture by explaining that the Náhuatl name *yoyotli* means rattles, to describe the ways the plant's seeds grow. It then invokes Hernández, the "wise physician of Felipe II" who tells us that the "ancient Mexicans used the milky juice" of "this tree to cure deafness and mange; and they attributed to the leaves, applied topically, the property of alleviating molar pain and resolving tumors; and, finally, they used the fruits in the treatment of ulcers."[53] The monograph then describes current medicinal uses, recounts the experiments Herrera undertook to determine the chemical composition, and proposes some additional potential uses that the author considered to be supported by modern science. It closes with suggestions for how the plant might be economically exploited for the benefit of Mexico. In this way, in place of the cold, abstracted plant descriptions idealized during the Enlightenment, Herrera's nationalist scientific approach rooted the plant deeply within Mexico's soil and history, and drew a direct lineage from the Aztecs through Hernández to the modern independent Mexican Republic. This format, with its inclusion of historical and ethnographic information, was the new standard within Mexican botany of the era.

Between 1869 and 1914, *La Naturaleza* published hundreds of such botanical monographs that refused to isolate a plant from its history or its place of origin. Each monograph marked the author as capable of contributing to universal science by nationalistically emphasizing the peculiarity of Mexican nature. As Mariano Bárcena explained, he and his fellow Mexican botanists followed in the footsteps of "Alzate, Mociño, Humboldt, Bonpland, Velázquez de León, Lallave [*sic*], Cervantes, del Rio, Bustamante, and many other wise men." They also paid tribute to "the great Linnaeus" along with Augustin Pyramus de Candolle, "Alfonso his son and Casimiro his grandson."[54] Such monographs cumulatively affirmed the republic as heir to its own ancient patrimony and to a distinctively national scientific tradition.

Starting in the mid-1880s, members of the SMHN bolstered this interpretation of their scientific lineage by using the pages of *La Naturaleza* to publish select historical scientific documents. An examination of the documents that the SMHN chose to publish is revealing. In 1884, for example, *La Naturaleza* republished Francisco Clavijero's long and detailed inventory of plants and animals of Mexico, originally published in 1780.

Clavijero started his account by invoking Hernández, the "Pliny of New Spain," but argued that, because Hernández was mainly interested in medicinals, he overlooked a lot of the native plants. Clavijero's text set out to address this gap.[55] It was no coincidence that this particular text drew the attention of SMHN members, who saw their own mission as an extension of the project initiated by Clavijero to document all native plants rather than just plants of medicinal use. In this way, they claimed Clavijero's list as something of a compass to guide their own effort to create a comprehensive nationalist inventory of Mexico's native flora.

La Naturaleza also republished Clavijero's famous refutation of the claims by the French naturalist the Count du Buffon (1707–1788) that flora and fauna in the New World were inferior to those of the Old World. It republished José Antonio de Alzate y Ramírez's 1788 charge that Linnaeus's method wrongheadedly ignored local observations about plants' medicinal uses, chemical properties, and growth habit.[56] And it republished botanical studies and speeches by José Mariano Mociño, with an eye for those that best affirmed the SMHN's own mission and that created the impression that Mociño was a protonationalist.[57]

An example of *La Naturaleza*'s selective use of Mociño is its 1884 republication of a speech he delivered back in 1802, in which he called for greater medical self-reliance by New Spain. In that speech Mociño encouraged doctors in Mexico to shun convoluted and flawed practices imported from Europe in favor of straightforward ones based on scientific experimentation with local medicinals. Here in New Spain, Mociño argued, "our land possesses almost all the medical substances" that were once well known among indigenous doctors, yet we stand "waiting for our pharmacies to be supplied by orders purchased from Europe." He compared New Spain to ancient Egypt, where despite growing wheat crops that fed the Roman empire the people were incapable of feeding themselves.[58] As steps toward a solution, he encouraged his peers to lessen their reliance on books written by Europeans and instead undertake careful study of the various plants that the expedition had begun to catalog.[59]

That same issue of *La Naturaleza* contained a 1798 speech in which Vicente Cervantes seemed to praise Hernández's employment of the Náhuatl system for naming plants. Using a particular variety of violet as an example, Cervantes demonstrated first how to use the Náhuatl name to understand the uses that indigenous people once had for plants, and second how to apply modern scientific methods to test how effectively the plants fulfill these curative claims. Though the members of the Royal Botanical Expedition rarely spoke positively of indigenous knowledge, the SMHN's selective curation of some of the rare speeches in which they did so became useful to scientific intellectuals of the late nineteenth century who sought to situate their efforts as part of a supposedly seamless lineage of nationalist engagement with native Mexican flora and indigenous heritage.

The following year, in 1885, *La Naturaleza* began publishing the biographies of Mexican naturalists, adding them one by one to the grand lineage of nationalist science. The series began with a biography of the Creole botanist Mociño, whom *La Naturaleza* now credited for most of the accomplishments of the Royal Botanical Expedition, including compilation of the flora mexicana.[60] It then published the biography of Cervantes, casting him as a Mexican patriot rather than a foreign agent of Spanish imperialism. After celebrating Mociño and Cervantes, the journal then moved to Miguel Bustamante, the Mexican student of Cervantes who succeeded his mentor as director of the botanical garden and head of the school of botany and medicine, and who wrote a botany textbook for Mexican students based on the teachings of Cervantes and his own observations about Mexican nature.[61] From this foundation, the pantheon of celebrated Mexican naturalists expanded as it grew closer to the authors' own time.

A few years into this biography series, the SMHN in 1887 brought to publication for the first time the Royal Botanical Expedition's 181-page list of all the plants it had documented in New Spain.[62] A few years later, in preparation for the presentation of this published list at the Chicago 1893 world's fair, members of the SMHN together with the Mexican government went a step further and eliminated the name "Royal Botanical Expedition" in favor of "The Sessé and Mociño Expedition," which they occasionally shortened further to simply "The Mociño Expedition," thus completing its transformation from an imperial endeavor into a nationalist project. Once again, what one called things mattered.

SACRALIZING MEXICO'S INDIGENOUS NATURE

Since most members of the SMHN were doctors, they gravitated toward medicinal plants, but they also lavished attention on plants that they knew had no medicinal or economic utility. This was the case, for example, with the cypress species known as ahuehuete, which after the revolution would be proclaimed the national tree of Mexico.[63] An even more intriguing example is the *árbol de manitas* (tree of little hands).

In his 1876 account, Mariano Bárcena informed readers that the Aztecs had esteemed this tree, which they called *macpalxochilcuahuitl,* because of its flower's similarity to a human hand. They introduced it to Hernández, who recorded his own detailed observations back in the 1570s. Then, in 1787, "the indígenas of Toluca" introduced it to Sessé and other members of the expedition by bringing them some flowers from the tree. Intrigued, Sessé had the Indians lead his team to the specimen that produced such flowers. They had the good fortune to find it at the season of the year when it displayed "all the organs necessary to study it to perfection" to draw it and categorize it within the Linnaean system (see TCHI 0172, figure 15 in chapter 4).

Sessé and his team hypothesized that the tree might have been planted in Toluca by an Aztec nobleman of the area rather than be native to the region. Cervantes later went in search of the place where the tree grew wild. He found it in Guatemala and suggested that maybe it was native to that area. But almost a century later, in 1871, Antonio Peñafiel (1839–1922, doctor, surgeon, and historian) was on a research expedition during which he found the árbol de manitas growing wild in Oaxaca. According to Bárcena, these and other expeditions had proven "that the arbol de manitas belongs" not to any particular region, but "to the flora mexicana" and thus was an essential element of the nation's cultural patrimony.[64] Bárcena used the árbol de manitas to again attest to an unbroken line from the Aztecs through Hernández and the Royal Botanical Expedition to botanists of his own era, affirming the indigeneity of the species and rooting it in Mexican soil, culture, and history. More importantly, he affirmed that the scientific work of the SMHN was about more than supporting economic development: it was a patriotic endeavor that anchored the republic deep in the flora, history, and soil of Mexico.

With close attention to the details of place and process, Bárcena recounted Sessé's devoted efforts to reproduce the árbol de manitas from seed and from twenty-five cuttings that he brought back to Mexico City from Toluca in 1787. Of these, only one cutting prospered. This was planted in the expedition's botanical garden in the viceregal palace (which later became the national palace, the symbolic center of the republic). Bárcena noted that, at the time of his writing in 1876, an árbol de manitas still stood in the garden of the national palace. But this twelve-meter-tall specimen, he was told, was not the original. Instead, it was an offspring that had sprouted from seed in the same spot as its parent around 1848, just as Mexico was recovering from the trauma of the Mexican-American War.

No botanical analysis of the time was complete without a dissection of flowers plucked fresh from the plant, followed by maceration of flower parts, leaves, stems, bark, and roots for chemical analysis. But an exception seems to have been made for the árbol de manitas. Bárcena noted that he collected only flowers that already had fallen to the ground. He dissected one of these fallen flowers and recorded his observations but seems to have protected the tree and even the dissected flower from the usual battery of indignities, instead performing a chemical analysis only of the soil in which the árbol de manitas was thriving.[65]

Bárcena provided the obligatory discussion on the medicinal uses of the plant, but this was uncharacteristically short and vague, guessing only that the tree's potential uses were probably similar to those of other kinds of Malvaceae—the vagueness of this statement becomes apparent when we consider that Malvaceae is a huge family that includes everything from hibiscus and cacao to cotton and durian, and even okra and hollyhock. One gets the sense that Bárcena would no sooner pulverize parts of the tree and its flowers for scientific analysis than he might the shroud of the Virgin of Guadalupe.

He noted that there were only three other known specimens of the árbol de manitas within the central valley—at San Francisco, Hacienda de los Morales, and near Tlalpan. Indigenous people regularly collected flowers from the specimen in Tlalpan to sell in the flower markets that dotted Mexico City. In their analyses of nonmedicinal native flowers, members of the SMHN often showed enthusiasm for the possibility of marketing them for their decorative value in homes, greenhouses, and manicured landscapes. They touted the potential of the *hauya elegans*, for example, as a decorative plant for use in bouquets. Since it was a native of the "flora mexicana" the *hauya elegans* supposedly made a better choice for Mexican gardens and bouquets than the various nonnative flowers that had been unjustly preferred "due to the whim of gardeners or the demands of fashion."[66] But, rather than present Indians' sale of flowers from the árbol de manitas in local flower markets as evidence of the potential to economically exploit the tree, Bárcena instead treated it as a restricted activity that was part of the tree's link to the nation's indigeneity. Each of these four specimens found in the central valley played a different role. While primacy was given to the specimen in the national palace, each, in its own way, as demonstrated by the one in Tlalpan, affirmed the Mexican Republic's claim as heir to a deep indigenous heritage rooted in nature.

CONCLUSION

Thanks to the efforts of the Sociedad Mexicana de Historia Natural and the Sociedad Mexicana de Geografía y Estadística, the two leading scientific organizations of the day, Mexicans made huge strides in the second half of the nineteenth century toward recovering "what had been lost."[67] These two organizations overlapped in their commitment to studying the national territory to bring prosperity to the republic while appropriating indigenous heritage. But there were crucial differences between the two organizations, including the SMHN's commitment to defining a distinctively Mexican scientific lineage and to promoting native species. Both organizations took a dim view of contemporary Indians, but the SMHN, with its commitment to things native, tended to be more generous regarding indigenous culture than was the SMGE. To members of the SMHN, it was enormously important where native plants were rooted, both physically in terms of place and figuratively in terms of history and culture.

Scientists of the SMHN emphasized their own role as heirs to a botanical and scientific genealogy that ran from the Aztecs and Francisco Hernández (whom they invoked as consistently as Renaissance botanists had invoked the Ancients), and through the Royal Botanical Expedition (which they rechristened the "Sessé and Mociño Expedition"). These scientists' efforts compellingly affirmed the republic as heir to an ancient patrimony rooted in indigeneity and to a distinctively Mexican national scientific tradition.

As their projects advanced, scientific intellectuals had to confront the question of how the parts of Mexican nature fit together as a whole, and to whose benefit. As I argue in the next chapter, in this search, they increasingly emphasized the connections between science, aesthetics, and medicine, recounted through the nationalist Romanticization of the natural world.

CHAPTER 7

NATURALIST-HEROES

Exploration, Landscape Painting, and Medicine in the Consolidation of Mexican Nature

Mexican scientific intellectuals of the late nineteenth century celebrated the opportunity to "study Nature, at the very place where Nature itself established its throne."[1] And, as shown in the previous chapter, they soon developed a pretty clear idea about how this celebration should proceed. Rather than emulate other countries, they championed a nationally distinctive relationship to their natural world, rooted in ideas of indigeneity and in Mexico's distinctive history of scientific endeavors. The more complicated questions were how to organize their disparate findings into a cohesive whole and whether their efforts were best achieved through dispassionate description, as their scientific predecessors had aspired to do during the Enlightenment, or through eyewitness narratives of discovery imbued with nationalist purpose. They also confronted the question of how to visualize Mexican nature in a way that bridged particular plants on the one hand and the vast space of the republic that cartographers were busily mapping on the other. And regarding written inventories of plant species, was it enough to simply compile lists and descriptions of native flora, or should these plants coalesce into a specifically Mexican style of medicine? This chapter addresses how Mexican scientific intellectuals and artists of the last part of the nineteenth century used narratives of discovery, modes of artistic depiction, and experimental medicinal practices to harness nature in service to their modernizing republic.

SCIENCE AND ROMANTICISM: NARRATIVES OF INTREPID EXPLORATION

A decade after Mexico achieved its independence, Pablo de la Llave, who had served as an assistant to Martín Sessé y Lacasta in the Royal Botanical Expedition, described the difficulty of exploring the country's botanical riches, even within his home state of Veracruz. "Botanizing in Europe," he claimed, considering the years he spent researching there during Mexico's independence wars, "presents no great difficulty. One walks through fields, plains, and abandoned land with small shrubs, and if one enters a forest, it is generally open and the trees are clear [of vines] with their shapes clearly visible. It is not so in the hot, humid regions of our land . . . Everything is covered with thick, impenetrable forests [*bosques macizos y obstruidos*], and the first thing a botanist must do is hire one or two huntsmen [monteros] as guides." To these dangers he added the risks presented by weather, swamps, illnesses, ticks, insects, spiders, snakes, and wild animals. "What an enormous difference between this situation and that of botanizing in Europe."[2] Though the purpose of his essay was to explain his proposed method for identifying trees by their bark and wood, it wandered into a personal narrative and adopted early hints of a Romantic tone in its descriptions of nature, emphasizing natural dangers and personal fortitude. When De la Llave first published his account in 1832, it stood out as an unusual blend of science and personal narrative, and as an early move toward Romantic descriptions of nature. Four decades later, in the 1870s, Mexican scientists ceased being so tentative. Instead, they embraced an overtly Romantic mode of describing their natural world. Regions where nature grew largely undisturbed, previously described as wastelands (*desiertos*) now emerged as sites of sublime beauty where one could commune with nature and discover the essence of the patria.

Botanical nationalists of the late nineteenth century claimed to draw inspiration from José Mociño, the Creole who rose to the leadership of the Royal Botanical Expedition. But Mociño was never the Romantic they pretended he was. Like his own mentors Martín Sessé and Vicente Cervantes, Mociño was narrowly committed to fitting Mexican plants into universalist Linnaean schemes. His descriptions of the landscape showed little interest in sentimentality or heroism, preferring a tone of dispassionate objectivity. This is evident, for example, in his 1793 account of his journey up Tuxtla Volcano in Veracruz. Rather than wax eloquent about the journey or the volcanic eruption, he provided matter-of-fact, unsentimental descriptions of each sound, sight, and event he encountered. His account coldly observed and measured each feature separately, in a manner that conformed to the Enlightenment practice of isolating particular phenomena as specimens, such that they never cohered as a landscape.[3] Naturalists of the late nineteenth century, by contrast, sought specifically to infuse their science with the kind of passion that Sessé and Mociño had avoided. Without compromising the integrity of their scientific method, they sought to make the natural world they witnessed cohere as a nationalist landscape that was distinctively Mexican.

Early German Romantic imagery had emphasized intense emotion through exaggerated displays of lightning, startled horses, rainstorms, and gathering clouds. By the second half of the nineteenth century, Romanticism had undergone a transformation, particularly in the Americas. It still emphasized intense emotion, but these emotions increasingly were positive and could be achieved by beholding the grandeur of God's creation or the sublime beauty of the natural world claimed by the nation.[4]

Romanticism in art and literature in the United States located the sublime in the country's "wilderness" frontier, where U.S. intellectuals and artists claimed to embrace the natural world unfiltered by civilization. As Henry David Thoreau proclaimed in his famous 1862 essay "Walking": "What business have I in the wood, if I am thinking of something out of the wood." Despite his claim to remove himself from society and history, with his mind on nothing other than his personal engagement with nature, Thoreau compared such a walk to a historical act of conquest: "For every walk is a sort of crusade . . . to go forth and conquer this Holy Land from the hands of Infidels." He even admitted that he and his walking companion fancied themselves "knights of a new, or rather an old, order,—not Equestrians or Chevaliers, not Ritters or Riders, but Walkers, a still more ancient and honorable class, I trust." He claimed that frequent walking of the kind he endorsed would toughen his character and thicken his skin, giving him "the tan and callus of experience" that brought him closer to the soul of the common man, whose weathered hands "are conversant with finer tissues of self-respect and heroism, whose touch thrills the heart, [more] than the languid fingers of idleness." During his walks in Massachusetts, Thoreau felt pulled westward by his instinctual search for freedom: "It is hard for me to believe that I shall find fair landscape of sufficient wilderness and freedom behind the eastern horizon." Only by heading "westward" into the wilderness do "I go free."[5]

We must not take the claims about wilderness by Thoreau and other U.S. naturalists at face value, particularly their boast that their encounters with nature were separate from their engagement with the political, social, and economic debates of their society. Rather, the very ways Thoreau and others constructed their understanding of wilderness, nature, and the western frontier were commentaries about their societies and the future of their nation at least as much as they were about nature, per se.

Like Thoreau, Mexican naturalists expressed their encounters with their natural world through first-person narratives. But rather than position their appreciation of nature as a flight from society, Mexicans emphasized its entanglements with contemporary state formation, ancient history, and modern science.

Luis Malanco (1831–1888, lawyer and politician) was not unusual in his embrace of this new Romantic spirit. In 1868, he waxed about the joys of "studying our land, following the course of our rivers, climbing our mountains, stopping to contemplate our lakes, finding under the layers of our atmosphere the breath of all climes" and "under the layers of our lands the life of all fruits; the richest minerals in our mountains, the most precious woods in our forests; almost all the useful animals in our fields, almost all the beneficial resources of our nature [*nuestra naturaleza*], all within horizons filled with light, among vegetation filled with exuberance, flowers filled with fragrance. Lift your eyes toward our sky, where you always find the most beautiful sun," take "in your hand the beautiful soil and" proclaim "that we live in Paradise on earth."[6] Rather than call for a march westward into new frontiers as did Thoreau, Malanco implored his countrymen to behold the natural wonders they already possessed but which they had only begun to study or appreciate.

This penchant for Romance flourished as a growing number of naturalists fanned out on expeditions to study Mexico's rural settings and then published accounts of their journeys. By the late 1880s, descriptions of expeditions brought the audience along for the journey. When the topographer and naturalist José Rovirosa (1849–1901) of the SMHN published an account of his ascension of mountains and exploration of jungles in southeastern Mexico in 1887, he unabashedly presented his scientific expedition as a heroic drama.[7] He told of daring ascents, dangerous encounters with snakes and other animals, and his emotional communion with breathtaking landscapes. When he passed through a cacao field, he saw it not just as a commercial crop, but as "a sublime spectacle" that "resembled a green tapestry of moss, limited to the N[orth] by the bluish vault [*bóveda*] that seems to rest on the plains of Tabasco."[8] His description of the vaulting of the sky over the plains evoked a cathedral, consistent with the sort of transcendentalist vocabulary favored at that time in the United States.[9] As he continued on his journey, Rovirosa found himself "in the heart of a virgin forest in every sense of the term. The grandeur of proportions, the dark green of the leaves of the New World vegetation, gave the landscape a grave and austere appearance unlike what one finds in the forests of Europe." To traverse this landscape, he had to hire Indians as guides and porters, and his inability to communicate with them deepened his sense of adventure even further: "When dealing with a land whose language and customs are almost completely unknown," discussions of science give way to pondering the mysteries of the human past and the interrelation among languages and people.[10] Rovirosa included an illustration of a mountain, which his Mexican contemporaries would have readily recognized as inspired by Alexander von Humboldt's famous drawing of Mount Chimborazo in Ecuador. This visual clue was fitting, since the tone embraced by Rovirosa and others seemed to draw on the Latin American writings of Humboldt, who, though a man of science, was driven by his senses and emotions as much as he was by instruments of measurement.[11]

Mary Louise Pratt has described how European imperialists of this era turned to natural history as a "means for narrating inland travel and exploration aimed not at the discovery of trade routes, but at territorial surveillance, appropriation of resources, and administrative control." Pratt focuses on the "naturalist-hero" as an agent of European imperialism whose "improving eye produces . . . 'empty' landscapes."[12] Mexican scientists adapted this "naturalist-hero" narrative to their efforts to study and claim indigenous nature in the name of the republic, but rather than envision the landscape as empty, as did Pratt's naturalists, they saw it as densely populated by history and layers of cultural signifiers. A central part of their nationalist mission, in fact, was to recover these truths about the deep entanglements between nature, nation, and indigeneity.

Mexicans' embrace of Romanticism was rooted in history and science rather than religion and wilderness as in the United States. In 1893, the physician and naturalist Manuel Villada (1841–1924) wrote an account of his exploration of the state of Veracruz, including an ancient forest that he estimated to be at least three and a half centuries old. Whereas a naturalist from the United States might have seen an old forest of this type in California as nature's cathedral and proof of his nation's rootedness in wilderness, Villada instead searched the ground for signs of ancient structures. He found "under the ancient shadows of the trees" traces of stone, clay, and lime that he speculated "could well have been temples, forts, and homes." These traces, he declared, were evidence of an ancient indigenous "race that knew well how to make use of the elements at its disposal." They also revealed that, when the "fatal hour arrived," this race "bequeathed its eternal name to posterity. Oh, such past heroic, epic-worthy deeds that, like sacred memories, deserve to be remembered!" These "silent ruins," evidence of this noble past, "hidden from sight, under a beautiful tapestry of foliage and flowers" created by "the tireless hand of nature," sat waiting to be discovered and claimed by explorers such as himself.[13]

The vegetation under whose shadow Villada discovered the ruins was not some dark jungle, as a foreign traveler might have perceived it, but specific tree and plant species with which the author demonstrated intimate familiarity. Every plant took on importance because Villada could announce its vernacular name as well as its Latinate binomial, and each place he visited took on importance because he could indicate its latitude and longitude, thus symbolically taking it into his, and the republic's, possession.

Villada did not mention plants, animals, or natural features unless he could name them. When he reached the coast and looked out into the ocean, he seemed at a loss about how to describe and possess a liquid space. He was saved from this potential crisis when he turned his eyes to the shore, where the land transmitted its historic associations to the waves, allowing Villada to realize the ocean's role as witness to Mexico's history: "How many memories are awakened by that mute witness's contemplation of so many historic events!"[14] Only after the ocean could be linked to the nation's deep history could Villada appreciate it and describe it as nature.

When not describing past civilizations or local plant species, Villada emphasized his own personal heroism and the difficulty of his journey.

The challenges and dangers he faced came not from wild nature, but from the lack of transportation infrastructure. While laying claim to the past and to nature, Villada welcomed a future in which increased intervention would bring improved infrastructure and development. This transformation, as he envisioned it, would be fueled by the capacity of men of science such as himself to know the name and potential use of each plant, the exact coordinates of every locality, and the history of each ancient indigenous civilization that could be claimed as Mexican patrimony.

Other authors celebrated nature by linking it to the contemporary built environment, rather than to that of the indigenous past. Juan Potts (1805–1876), for example, wrote for the SMGE's *Boletín* about his intimate engagement with the landscape, plants, and animals of Chihuahua. Though born in England, Potts spent his life in Mexico and worked as mint director (*director de la casa de moneda*) in Chihuahua. According to some accounts, he had even helped provision weapons to the Mexican army defending the city against U.S. troops during the Mexican-American War. By the last part of the century, Potts had become a powerful figure in Chihuahua politics, but he above all delighted in his experience as a naturalist.[15] His article starts by situating the reader within the city of Chihuahua, declaring that "few cities in the Republic and very few in the world are located in such a picturesque and beautiful location. Seen in the morning from the road that leads to Paso del Norte, . . . surrounded by a fantastic mountain range, the city appears to be framed as a painting worthy of being captured by the brush of the greatest of landscape artists."[16] Rather than simply invite the reader to behold the picturesque landscape from a distance, he takes us on a journey into it.

Like many environmentalists then and now, and tracing back to Pablo de la Llave, much of Potts's connection to nature was tied to his experience as an outdoorsman and a hunter. Unlike his counterparts in Australia and Argentina of that era, who felt that their native faunas were unfit for manly hunting, Potts celebrated the native animals.[17] He praised the opportunities to hunt "black and brown bear" in Chihuahua's "high mountains, along with wild boar, several varieties of wolf, and bighorn sheep. Antelope, black-tailed deer, and American red deer abound in the plains and mountains. Hare, rabbit, partridge, quail, snipe, ducks, and geese and infinite other waterfowl." He similarly admired the variety of Chihuahua's native flora, most of which he had studied firsthand and some of which he even had the honor of discovering, naming, and introducing to Europe and the United States.[18] Though Potts may have been born in England, he spent his life in Chihuahua, and his view of the region's natural world blended science, adventure, and the sublime in a Mexican idiom.

Even the doctor José Ramírez (1852–1904, physician and son of "El Nigromante" Ignacio Ramírez), who claimed to shun heroics and Romanticism, could not resist waxing poetic when he ascended Popocatépetl and looked out across the landscape, down into the volcano's crater, and up to the clear sky.[19] Rather than standing in contradiction to one another, scientific rigor and aesthetic means of conceiving the nation unfolded hand in hand during the latter part of the nineteenth century. This was most evident in the art of José María Velasco, who saw Mexico's landscape as the divine embodiment of the republic, which could best be comprehended through the marriage of art and science.

PICTURING MEXICAN NATURE: THE LANDSCAPES OF JOSÉ MARÍA VELASCO

José María Velasco (1840–1912) is widely regarded as one of Mexico's greatest artists. Less commonly known is that, in addition to being a landscape painter, he was a naturalist, a botanical illustrator, and even an executive member of the Sociedad Mexicana de Historia Natural. And, as art historian James Oles has noted, more than any other artist of his era, Velasco combined "objective science and poetic invention in the service of nationalism."[20]

When Velasco was a youth, he and his family left their hometown of Temascaltepec, Estado de Mexico, and settled in Mexico City, near the old Potrero de Atlampa site of the Royal Botanical Garden. At the age of eighteen, he enrolled at the Academia de San Carlos de Arte, where he became enthusiastic about the teachings of Eugenio Landesio, an Italian artist who had been recruited just three years earlier as the Academia's first professor of landscape painting.

It was Landesio who taught Velasco to approach landscape painting from a scientific perspective. The master first had students copy his own drawings and lithographs of particular plant species, then released them to create plant studies from nature. Students then practiced the fundamentals of geometry and perspective until they were finally allowed to assemble their own landscapes. While enrolled at the art academy, Velasco also attended the national school of medicine, where he studied physics, zoology, and botany.[21] He united these scientific and artistic interests to develop his own style of landscape painting in which he meticulously depicted the characteristic details of each particular plant species within sweeping, historically infused panoramas. Velasco's elevation of local plants as elements within a coherent nationalist epic was recognized at the time as a novel convergence of art and natural history.[22]

Velasco's earliest successes came from paintings of urban landscapes rather than of nature. At that time, colonial-era buildings within Mexico City were being demolished to be replaced with modern architecture in the French style. Velasco took advantage of these settings to capture a sense of nostalgia for a crumbling past, while also expressing hope for a modern future, a tension that would remain central to his paintings as he began to look past the urban center into the vast countryside of central and southern Mexico.

Velasco extended his interest beyond the city thanks to a grant he won to spend several weeks in isolated natural settings painting en plein air.[23] From that moment forward, he embraced opportunities to paint in rural locations and took to heart the advice of his mentor Landesio, not just in the meticulous detailing of individual plants, but also in how he understood the craft of depicting a landscape. In a guide he wrote for his students, Landesio explained that the artist must paint a location in a manner that brings dignity to "the history of that place, giving a clear idea of the site, not materially and servilely copied, but rather liberally and poetically, transporting the observer" as much as possible to the particular place and time. The painter must not omit "the geological characteristics of the land, nor the structure and style of the buildings." Neither should he omit details regarding "the indigenous plants and their characteristics." At the same time, the artist should feel free to shape the scene by "adding or removing, modifying this or that other object in whatever ways seem appropriate to the understanding of the subject of the place, to enhance its elegance or grandness" and to "stimulate and direct [*disponer*] the viewer's soul, whether toward melancholy or pleasure, the terrifying or the solemn," so that "the moral and spiritual aspects merge with the objects and peculiarities of the site, the atmosphere, the play of light and shade, into an intimate and harmonious union."[24] Velasco embraced Landesio's lessons regarding depiction of individual plants, narration of landscapes, and the need for rigorously planned composition, and refined these to a level beyond even what his mentor could have hoped for.

El cabrio de San Ángel/La Hormiga (*The Goatherd of San Angel/The Hormiga Factory*), 1863 (figure 25), offers an example of how José María Velasco already was masterfully putting this advice into practice when he was just twenty-two years old. The viewer's eye is first pulled to a cluster of trees in the top left, then it glides downward along a barren slope toward what at first looks like a natural waterfall, but which is revealed to be the outlet from a dam. The water drops into a deep gorge that cuts across the canvas diagonally. Observation of dramatic forms gives way to inspection of details. This reveals a tangled maguey on the lower left, beyond which a traditional goatherder is anxiously directing errant members of his herd away from the cliff. Across the abyss rises a modern factory—La Hormiga, which at that time was the country's second-largest textile manufacturer—with its smokestack and fumes harmoniously blending into the treetops and the clouds. A guard sits perched almost imperceptibly atop the factory roof, watching the goatherder. The seemingly unbridgeable chasm between Mexico's past and its future leaves the traditional goatherder with no option other than retreat. The only things that seem to unite the peasant past and the industrialized future is a shared natural world, which Velasco paints with meticulous attention to each plant species. The result is a brilliant merger of scientific realism, history painting, and Romanticism.

Velasco was working at a time when artists in multiple countries partook of a shared zeitgeist in which they each turned to landscape painting in search of their own distinctively national artistic tradition. A comparison to works produced by artists from the United States reveals certain commonalities but also instructive differences. Most notably, Velasco embraced a more scientific view of nature than did his contemporaries in Europe or the United States. And rather than fixate on the frontier or wilderness, as did members of the U.S. Hudson River School, Velasco endowed nature with the weight of human history and saw in it the promise for the nation's industrialized future.[25] We can see this, for example, by comparing *El cabrio de San Ángel* to Albert Bierstadt's famous *Valley of the Yosemite* (1864), a U.S. Romantic painting completed a year later. Both landscapes employ strikingly similar compositions and uses of light and

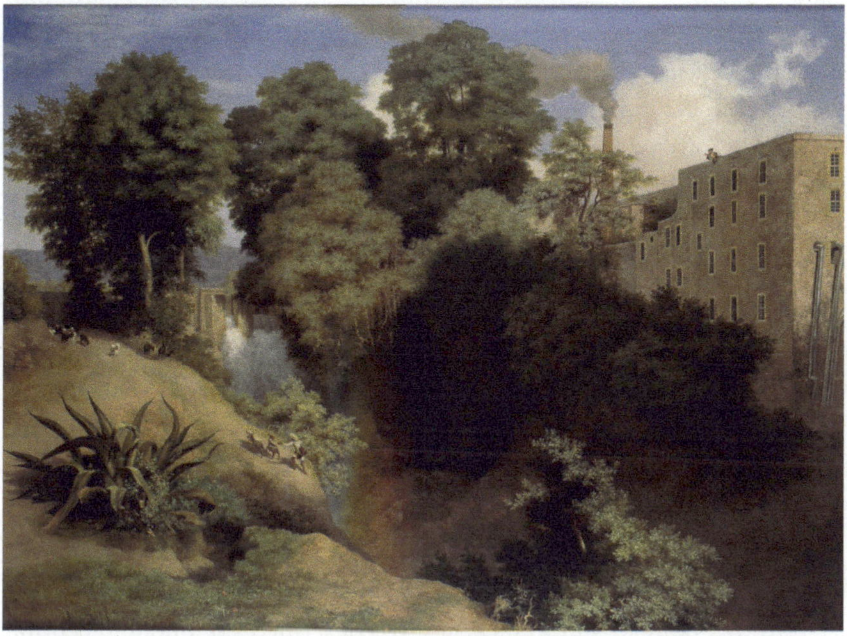

FIGURE 25 José María Velasco, *El cabrio de San Ángel/La Hormiga*, 1863. Museo Nacional de Arte, Mexico City. Courtesy of the Instituto Nacional de Bellas Artes y Literatura, 2024.

color, but while Bierstadt constructs nature as wilderness, unmarred by the presence of modernity, Velasco portrays human activity and the nation's nature as tightly interwoven.

When Velasco graduated from the Academia de San Carlos in 1868, the Academy promptly hired him as one of its professors. That same year, he began work as a botanical illustrator for the *Flora del Valle de México* and was inaugurated as a member of the SMHN.[26] In the years ahead, he contributed scientific images for numerous articles published in the SMHN's journal *La Naturaleza*. He also published his own monographs of various plant species, along with multiple studies of the axolotl (an unusual Mexican salamander). In addition to his work as a teacher, a naturalist, and a prolific landscape painter, Velasco worked for the Museo Nacional drawing archaeological objects and sites that were considered national patrimony. Through these various activities, he became a close friend to Mexico's leading naturalists, particularly Fernando Altamirano, who eventually became his in-law. His scientific colleagues held such respect for Velasco that they elected him to several leadership roles within the SMHN, including vice president.[27]

Velasco traveled frequently across the country as an artist and as a natural historian, bearing witness to the diversity of Mexico's nature. When seeking out sites for his landscape paintings, he brought along his botanical toolkit that included tweezers, dissection knives, a set of magnifying glasses, collection jars, and a notebook in which to sketch plant species that grew in that setting. He also collected leaves and petals from the plants he observed and pressed them between the pages of his notebook, like a personal herbarium.[28] Everywhere he went, Velasco admired both the sweeping vistas and the individual plants and animals that populated each location.

The artist's first international trip came in 1889 when, at the age of forty-nine, he accompanied his canvases to Paris for display at the Exposition Universelle, stopping over in Cuba and Spain on the way. In a letter to his family from France, Velasco described the public reception of his paintings as very positive, stating that "everyone praises them, and the [depictions] of Mexican nature draw much attention."[29] From Paris, Velasco continued to Austria, Switzerland, Germany, Italy, and England, visiting the great masterworks of European art, including those by Peter Paul Rubens, Anthony Van Dyck, Rembrandt, and Raphael. Four years later, in 1893, Velasco accompanied his art to the Universal Colombian Exposition in Chicago, the same event where the SMHN first presented to the world the published findings of the "Sessé and Mociño Expedition."

The art Velasco saw in Europe and the United States convinced him that Mexican artists should expose themselves more to cosmopolitan currents. At the same time, he felt that he and his countrymen had surpassed the achievements of their European contemporaries in the category of landscape painting. Even the French landscape painters, despite the beauty of their compositions, fell short because they lacked a convincing connection to the natural world they depicted. Among other issues, they "don't make it possible to recognize the particular plants [*no se puedan caracterizar las plantas*]" found in those landscapes. Mexican painters, in his view, surpassed their European counterparts, even though depiction of the Mexican landscape was more challenging owing to the density and diversity of its vegetation. The importance of capturing specificity was crucial, he argued, even in sweeping vistas (*lontananzas*), because, though "the mountains may be far away, they are full of details" that are important to the locality.[30]

Velasco created numerous landscapes that put into practice what he had in mind. His works were masterfully composed while paying close attention to the geographic, historical, and botanical specifics of each place. Among his many outstanding examples is *Cañada de Metlac*, circa 1896, depicting a celebrated section of the Mexico City–Veracruz railroad line that ran through the Barranca de Metlac/Metlac Ravine (figure 26 and map 2).

The effort to connect the capital to the port in Veracruz via rail began in 1837 but had shown slow progress. By 1860, when the United States boasted over thirty thousand miles of track, Mexico had barely one hundred fifty miles. Between 1864 and 1866, under Emperor Maximilian, British engineers working for the Ferrocarril Imperial Mexicano made halting progress on sections of the line between Mexico City and Veracruz until the company went bankrupt as a result of the expense and engineering challenges of laying track across this difficult terrain. Under Juárez's restored republic, the government rescued the company, which it renamed Ferrocarril Mexicano. The Metlac Ravine was one the most difficult sections of the line, and its completion was announced with fanfare in January 1873.[31] Even with completion of the Mexico City–Veracruz line, Mexico had a mere 355 miles of track, of which 114 still relied on mules rather than steam engines to pull cargo. By the early 1880s, Mexico's future finally became more visible as the rail network expanded to over three thousand miles, now fully served by locomotives rather than draft animals. This doubled in the next decade, then doubled again by the time the Mexican Revolution broke out in 1910.[32]

Mexican leaders and intellectuals recognized completion of the line through Metlac Ravine as the symbolic turning point. The government had at last overcome a long-standing and daunting engineering challenge, and in doing so finally connected the capital to its main port. To Velasco and other contemporaries, Metlac symbolized the nation's reconciliation of its intractable past with its modern, industrialized future.

Velasco was fascinated by the railroad in Metlac Ravine and painted it many times from various vantage points beginning in 1881. Whereas most of his earlier versions focused on the bridge that spanned the gorge, his circa 1896 portrayal (figure 26) shows a locomotive, occupying much

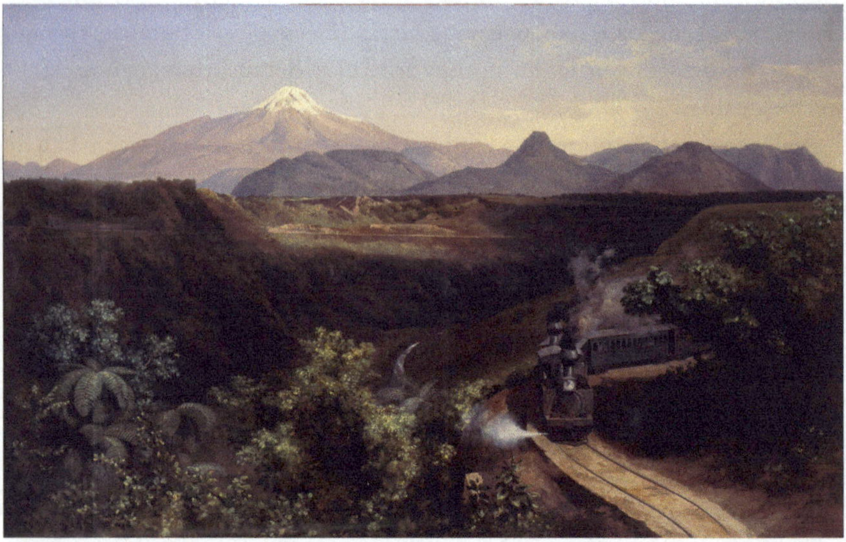

FIGURE 26 José María Velasco, *Cañada de Metlac*, circa 1896. Museo Nacional de Arte, Mexico City. Courtesy of the Instituto Nacional de Bellas Artes y Literatura, 2024.

of the lower right, coming around a bend along a steep cliff, about to skirt just right of the viewer (from there it would have come round behind us to cross the bridge). The ravine sweeps from the upper right down toward the center of the painting, with a river barely visible in its depth. Most of the painting is in shadow, except for the upper left, where the snow-capped Pico de Orizaba (Citlaltépetl) rises in the sunny distance. In the left midfield sits a barely discernible continuation of the train track that disappears into a tunnel before reemerging in the sunlit distance. The lower-left corner is filled with lush greenery, with each plant meticulously depicted.

To appreciate some of what Velasco accomplished in *Cañada de Metlac* we can compare it to other well-known portrayals of similar landscapes from the era, such as those by U.S. artist Frederick Edwin Church, whose iconic depictions of the Andes exaggerated the tropicality of the Latin American landscape, showing exuberant overgrowth of unidentifiable plants, with occasional palm trees rising from the jungle (a cliché that U.S. and European artists frequently employed to mark Latin America as tropical).[33] Velasco had no interest in such trite formulas. Instead of an unruly and exaggeratedly fecund jungle, he painted every plant as a recognizable species that grows in that location. Far from a mysterious landscape overwhelmed by tropical nature, Metlac Ravine emerges as a specific site where technology and nature coexist in harmony, thereby defining a vision of Mexico that is nostalgic yet forward-looking, headed toward a nationally distinctive modernity.

In 1901, Velasco severely injured his leg when the streetcar he was riding collided with a locomotive. Though his two good friends and fellow SMHN members, the physicians Fernando Altamirano and Manuel Villada, provided him with the best medical care available, the sixty-one-year-old now found it difficult to travel to rural locations for open-air painting. As a result, he based most of his future work on the many photographs and studies he had produced over the previous decades, including detailed sketches of plants, people, and geographic details. He also devoted more time to commissions from the Museo Nacional to create illustrations of archaeological objects and imagined landscapes of Mexico's geological past.[34]

Velasco's *El cabrio de San Ángel/La Hormiga* and *Cañada de Metlac*, along with his numerous other landscape paintings, beautifully merged science and Romance, while visually testifying to the spatial integrity of Mexican nature. But Velasco was not alone in giving visual form to these discourses. Moreover, as he and other landscape painters incorporated science into their Romantic landscape paintings, scientists reciprocated by incorporating Romantic views of botanical nationalism into their scientific illustrations.

BOTANICAL NATIONALISM IN ANTONIO GARCÍA CUBAS'S CARTOGRAPHIC LANDSCAPES

When people think of Antonio García Cubas (1832–1912), they usually have in mind his national maps of Mexico. But he also should be remembered for the artistic images that filled the borders around his maps. His landmark *Carta General* of 1858, for instance, contains two landscape prints hovering above the map. One is a composite image of at least four well-known Mayan archaeological pyramids, crumbling as they are consumed by time and jungle. The second is what historian Raymond Craib describes as an "impossible" landscape that meshes seven well-known topographical features, geographically distant from one another along the route between Mexico City and the port of Veracruz, as though they can be viewed in a single panorama.[35] Five years later, García Cubas published another *Carta General* that, in place of invented composite landscapes, depicted specific locations: one in Puebla and the other just outside Mexico City. After Juárez's restoration of the republic, García Cubas borrowed from the approach to landscape promoted by painters like Velasco and printmakers such as Casimiro Castro (1826–1889; see chapter 8) to guide his attention to particular locations along with specific plants and ecological zones.

This shift is evident in his *Atlas pintoresco é histórico de los Estados Unidos Mexicanos* (*Picturesque and Historical Atlas of the United States of Mexico*) produced in 1885. The publication contains thirteen themed full-color maps of the country, each surrounded by images of locations related to the subject of the map.[36] The communications map, for example, incorporates color prints of important ports, such as Acapulco, and recently completed railroad bridges including the one spanning Metlac Ravine. The Valley of Mexico map contains images of Chapultepec as well as the same Hormiga textile factory in San Ángel that Velasco had incorporated into his *El cabrio de San Ángel*. And the orographic and hydrological maps are surrounded by images of sites such as Popocatépetl, Pico de Orizaba, Catarata de Juanacatlán, and Rio Balsas. While the images in these maps are more naturalistic than those that surrounded García Cubas's earlier maps, each nonetheless is simplified into a recognizable nationalist icon with just enough information to make it recognizable.

Within this atlas, the agricultural map functions somewhat differently and more interestingly than the others (see detail in figure 27). Some images on this map depict well-known haciendas, but others focus on particular kinds of ecosystems rather than on iconic locations. For example, the image titled *La cosecha* (*The Harvest*), centered above the map (see the top-left corner of the image), shows a long row of peons lined up with their baskets and machetes, ready to attack the dense sugarcane field. Behind them is a line of mounted overseers eyeing the artist, and beyond the workers rise distant mountains. In this manner, the image captures a particular ecological context for sugar production, as well as the exploitative political and economic relations that structure this agricultural system—notably, the exploitation of peons appears as mere observation rather than as social critique. Just above the map title, to the right of

La cosecha, is *El riego* (*The Irrigation*), which similarly depicts the exploitation of peons within a particular kind of ecological setting. To its right is *Soapayuca Hacienda de Pulque-México*. Rather than show the labor of man, this landscape depicts the work of nature, as one of the maguey plants closest to the viewer enters into full bloom, its stalk topped by bright red and yellow flowers (readily observed in the original image, though perhaps hard to discern in this reproduction).

Even more notable is the image below the maguey plantation, to the right of the map. Though it accompanies an agricultural map, many of the plants in this image have little to no agricultural utility. Each is labeled

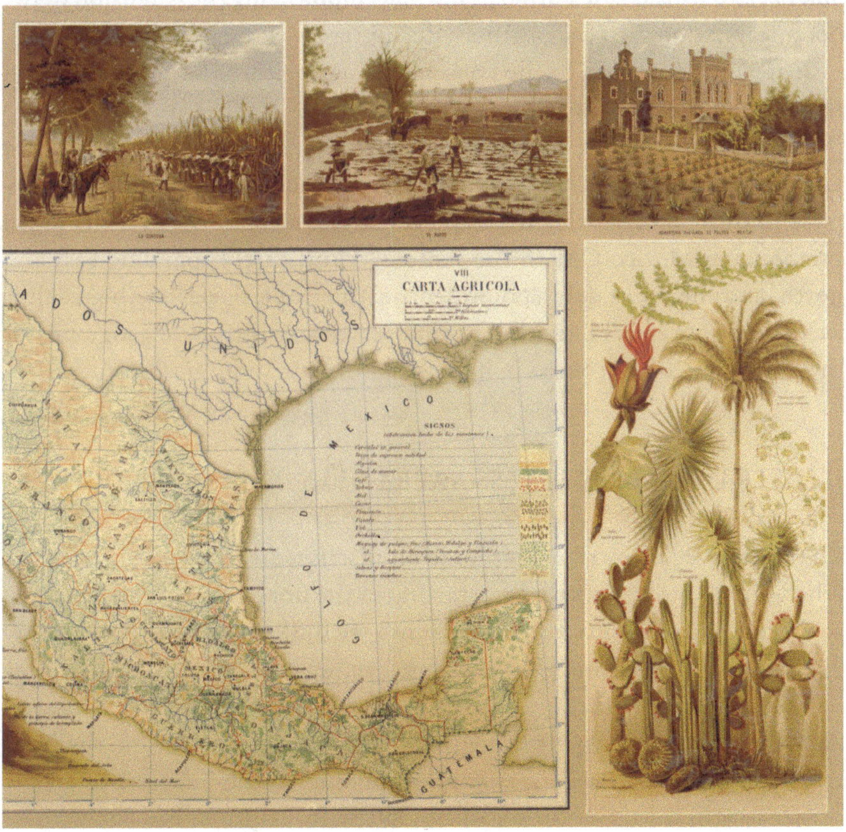

FIGURE 27 Unattributed artist, detail from Antonio García Cubas, "Carta agrícola," *Atlas pintoresco é histórico de los Estados Unidos Mexicanos* (Mexico, Debray Sucesores, 1885). Courtesy of the U.S. Library of Congress. https://www.loc.gov/item/2008621671/.

with its common name, along with its Linnaean binomial. Among the plants is the famous *árbol de manitas* depicted in TCHI 0172 (figure 15). As discussed in chapter 6, the flowers of this tree had captured the imagination of Aztec nobility, Hernández, the Royal Botanical Expedition, and then nineteenth-century nationalist botanists, thereby emerging as a symbol of continuity that established the modern patria as heir to a botanical tradition that traced back to time immemorial. Similarly, there is the nopal, whose importance as a symbol for the founding of Tenochtitlán would not have been lost on the viewer. Other plants such as the biznaga cactus and yucca (both of which attracted attention from the Sessé and Mociño Expedition) and the organ cactus (a favorite subject of Velasco) similarly link back to Mexico's scientific past while emphasizing the extension of the botanical nation into the northern desert territories.

Instead of speaking to the commercial agriculture theme of the map, this image displays scientific knowledge about plants that are important to the history of Mexican botany. And the arrangement of the plants is not arbitrary; it is a botanical geography of the republic, with the árbol de manitas to the left suggesting the importance of southern Mexico, the nopal symbolizing Mexico City in the center, various cacti expanding east and north, and the *palma de coyol* representing the tropical coasts. The result is a cluster of botanical illustrations that serves as a historical geography of the Mexican nation, rooted in nature, history, and botanical science.

ILLUSTRATING EYEWITNESS NARRATIVES OF EXPLORATION

The narratives of intrepid exploration discussed earlier in this chapter relied on the written word. But starting in the final decade of the century, scientists supplemented their accounts with illustrations. Their incorporation of images was aided by improvements in printing technology and likely inspired by the paintings by José María Velasco, the atlases by Antonio García Cubas, the lithographs by Casimiro Castro, and the illustrated accounts by Alexander von Humboldt. These images helped the reader to join the adventuring author in the Romance of bearing witness to Mexican nature.

An example of this is a study of the hot springs of Michoacán that Manuel Villada published in *La Naturaleza* in 1891. The author begins by

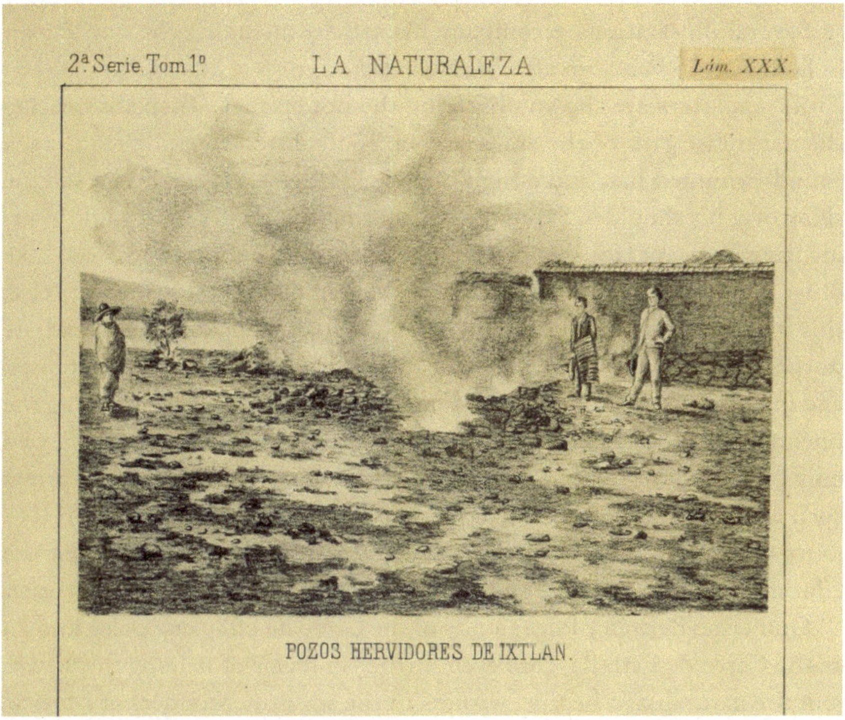

FIGURE 28 Unattributed artist, *Pozos hervidores de Ixtlán*, from Manuel Villada, "Apuntes de Geología y de Botánica Relativos a México, Erupciones de agua caliente en el Estado de Michoacán," *La Naturaleza* 2a serie, tomo 1 (1891), plate XXX following page 432, courtesy of the American Museum of Natural History.

establishing himself as an explorer and eyewitness to the features he sets out to study: "In the years of 1887 and 88, I had the occasion to observe the eruptive phenomena to which I refer, in two different locations in Michoacán."[37] Throughout the article, the author serves as the readers' guide, bringing us along step-by-step as he walks the landscape, helping us join him in bearing witness to Mexican nature. At each stop, Villada tells us the longitude and latitude and shares various geographic and hydrological observations. He also lists each of the plant species at the site and speculates about what these plants might reveal about the site's connection to broader ecological zones. In these ways, Villada allows the readers to feel as though we are joining him in his exploration of the landscape, nature, and particular scientific phenomena.

Several illustrations accompany his article, including the one shown in figure 28. In *Pozos hervidores de Ixtlán* (*Hot Springs of Ixtlán*), the two scientific explorers are shown observing the hot springs. They are wearing the naturalist garb of the era—leather boots, fitted sporting trousers, a round-brimmed hat, and a high-quality sarape—and one has a satchel slung over his shoulder. Standing at a cautious distance, a local guide wearing baggy peasant trousers and wrapped in a simple sarape looks on. Like other illustrations that increasingly accompanied natural history articles, this image does not focus on any particular scientific detail. Instead, its purpose seems to be to validate the eyewitness status of the naturalists, and invite the viewer to partake in the act of exploring and witnessing the phenomena. Whereas the paintings by Velasco might have been viewed only by the visitors to a museum, world's fair, or private collection, prints such as this had the potential to reach a broad audience via the SMHN's journal *La Naturaleza*. By reading the account and viewing the images, the educated public could become part of this nationalist scientific enterprise.

Guillermo Beltrán y Puga's study of the Cerro de Ixtapalapa, also known as the Cerro de Estrella, goes even further in terms of helping the viewer join the naturalist in bearing witness to the scientific wonders of Mexico. Puga (possibly 1863–1939, a geological engineer who went on to help plan the famous Angel of Independence on Paseo de la Reforma) began his account by inviting the reader to take a Romantic view of the site: "The picturesque town of Ixtapalapa is well-known to the inhabitants of this capital. Comprised mostly of a group of gracious and fertile chinampas, it presents a cheerful appearance, while stirring memories of the primitive way our Aztec ancestors used to build their homes on floating gardens."[38] Like Villada, Puga then takes the viewer on a walking tour in which he frequently stops and invites us to partake in observations that flow seamlessly from the aesthetic to the scientific. An accompanying illustration, (figure 29), which purports simply to show the geologic structure of the hill, instead does much more than that.

The geological drawing within the illustration occupies less than half the page, while the rest is taken up by two prints. One, titled *Roca volada en la cima del cerro de la Estrella* (*Jutting Rock at the Top of the Star Hill*) shows an explorer standing on a jutting cliff, looking out over the valley floor. Wearing a standard expeditionary outfit, with a sarape draped over his shoulders like a cape, the naturalist takes in the view from high on the

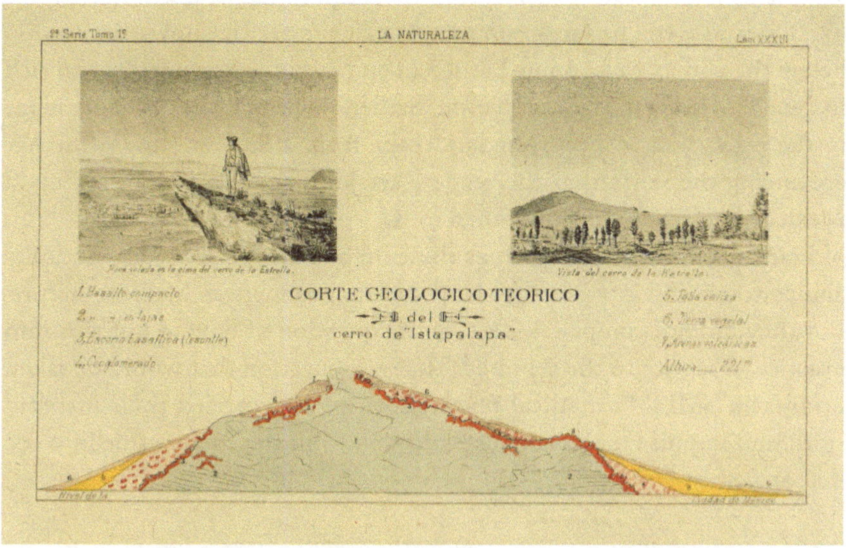

FIGURE 29 Unattributed artist, *Corte geológico teórico del cerro de 'istapalapa'*, *La Naturaleza* 2a serie, tomo 1, 1891, Plate XXXIII, following page 492, courtesy of the American Museum of Natural History.

cliff, while giving the reader the sense that we are bearing witness along with him. This image seems to be an intentional reference to the famous painting *Wanderer Above the Sea of Fog* (c. 1818) by the German artist Caspar David Friedrich, which at that time already was considered a quintessentially Romantic work of art. In Friedrich's painting, an adventuring naturalist dressed in hunting boots and a hiking jacket, cane in hand, boldly stands on a rocky precipice looking out over a craggy landscape shrouded in mist. But in contrast to the German masterpiece, the landscape in front of Puga's protagonist is not shrouded by fog but drenched in sunlight and crisp in detail. The print also changes the angle as compared to Friedrich's painting, to make it more inviting for the viewer. In doing so, Puga's image employs a device used by other Romantic nationalist landscape painters of the Americas, including the U.S. painter Asher Brown Durand, whose *Kindred Spirits* (1849) shows two men on a jutting cliff contemplating the wilderness beyond. A similar device had been employed in 1877 by Casimiro Castro in his well-known print *Infiernillo* (see figure 30). Rather than portray the landscape as foreboding like Friedrich's, or as wilderness like Durand's, the image accompanying Puga's article shows the terrain below the cliff as welcoming, bathed in sun, occupied by people, and rich in detail. To draw the viewer in even further, Puga includes a second image to the right, which shows the landscape from a different angle, further enhancing the sense that the readers are joining the explorer, taking in Mexico's landscape with their own eyes.

Narratives and images such as these helped the viewer to aesthetically and conceptually integrate the land and its products as Mexican nature. But they did not completely resolve the question of how to put this into practice in service to the republic. That is a question that would be taken up by the SMHN's medical researchers as they carried their intrepid spirit and impulse for nationalist cohesion from the countryside into the doctor's office.

CREATION OF THE INSTITUTO MÉDICO NACIONAL AND THE CONTINUED PURSUIT OF AN INVENTORY OF MEXICAN NATURE

In 1892, the SMHN updated its mission statement to affirm that, while it sought to add to world science, its primary commitment was to contribute to the "biological knowledge of Mexico" and toward the development of a distinctively Mexican approach to medicine.[39] The Secretaría de Fomento was sufficiently impressed by the progress that the SMHN was making in its study of indigenous medicinal plants that it recruited a dream team of leading doctors and botanists from within the organization to undertake a special project.

Their task was to assemble two thousand five hundred native specimens for the Exposition Universelle de 1889 in Paris, which historian Mauricio Tenorio-Trillo has described as "the greatest [world's] fair of the nineteenth century."[40] This was the same event, incidentally, at which José María Velasco first garnered international acclaim for his landscapes. The team included José Ramírez, Alfonso Herrera, and Francisco Río de la Loza (1856–1901), who was a physician and son of Leopoldo Río de la Loza (1807–1876, a physician, chemist, and public health advocate who had drafted Veracruz's forestry laws after studying under Vicente Cervantes at the botanical garden and its school of medicine). To lead the project, the Secretaría hired Fernando Altamirano (1848–1908), an accomplished doctor and botanist, as well as in-law to José María Velasco and nephew of Manuel Altamirano—who, like Leopoldo Río de la Loza, had studied under Cervantes.

The exhibition of native Mexican medicinal plants at the Paris world's fair was such a success that the national legislature allocated 30,000 pesos for the same team to create the Instituto Médico Nacional (IMN) as an "establishment dedicated exclusively to the scientific study of our medicinal products."[41] The IMN was formally inaugurated on August 14, 1890, under the direction of Fernando Altamirano, assisted by his fellow medical researchers and members of the SMHN José Ramírez, Alfonso Herrera, Francisco Río de la Loza, and Secundino Sosa (1857–1901).

Many members of this team also had been central players in the search through European archives for the lost manuscripts of Sessé and Mociño. This overlap of activities led to a second triumph on the international stage, when they first publicly presented the previously discussed long-lost findings of the "Sessé and Mociño Expedition" at the 1893 Columbian Exposition in Chicago. This was the same world's fair at which Frederick Jackson Turner issued his famous declaration of American (by which he meant the United States) exceptionalism. According to Turner, every U.S. pioneer who ventured westward had a hand in recreating civilization from the frontier wilderness, each time forging an ever more distinctive American democracy.[42] Within their own pavilion, members of the SMHN, operating under the auspices of the Instituto Médico Nacional and backed by the Secretaría de Fomento, proudly presented the Sessé and Mociño Expedition's "Plantas de Nueva España" as part of their own nation's special relationship to its natural world.[43] As previously mentioned, this 1893 publication and public presentation was a formal, international assertion of Mexico's rights over nature within its territory, as well as all the knowledge produced about that nature from the time of the Aztecs to their own day.[44]

A few years later, in a fiery 1897 speech delivered at a meeting of the SMHN, Gabriel Alcocer made a case for taking the next step toward an updated and comprehensive Mexican flora that could serve the medicinal, economic, and nationalist needs of the patria. He argued that, from the ashes of the independence and civil wars, foreign invasions, and sacrifices made by many citizens, Mexico at last had emerged as a land of order, civilization, and economic progress. "But that is not enough." Just as Mexico had created a map of the republic, "we, in the name of the Sociedad we represent [i.e., the SMHN]" call on the country to build on the recently recovered Sessé and Mociño findings by working toward an updated comprehensive "catalog of all the plants produced in [Mexico's] soil."

Once Mexico completes the "botanical exploration of all the regions of the country and assembles the Flora Nacional," the benefits of this "Geografía Botánica" would be not just scientific, but also practical. Knowledge of "all the plant populations that live in our extensive territory" will not just result in pages "full of Greek and Latin terms" but also help us to know how to exploit national botanical resources for the benefit of the country.[45] The purpose of such an inventory, therefore, was not to fulfill "mere scientific curiosity, or vainglorious erudition," but rather to empower Mexico to use its natural possessions to support its own needs, and even develop new products for export.[46]

"All the educated peoples" of the world, according to Alcocer, "have understood the necessity, the capital importance of getting to know their resources" that are "at their disposal." To truly join the ranks of the world's most civilized nations, Mexico, too, must know its own resources. Mexico had a head start compared to other countries in the Americas thanks to the pioneering work of the Hernández expedition, then, two centuries later, the Sessé and Mociño Expedition, but recently "it has fallen a few steps behind."

More troubling, Alcocer argued, was that foreigners from Europe and the United States had surpassed Mexicans in their knowledge about what botanical resources were available within the Mexican territory. He drew an explicit link to foreign military invasion: "As we can see," foreign governments, most notably those of the United States and France "upon unjustly invading us" and subjecting us to "terrible and barbarous" warfare, made sure to also exploit us for their scientific interests, sending "their doctors, surgeons, and pharmacists" to Mexico along with their military forces, with instructions to study our plants and assemble collections that they then delivered "to the scientific centers and scientists of their [own] countries."[47]

In a grand gesture of postcolonial recuperation, Alcocer argued that now that these foreigners had begun to publish their findings, Mexicans could look to their work "as a guide that will greatly facilitate" our own efforts to compile our flora mexicana. Alcocer also urged his colleagues to draw upon the existing labors and methods of the Comisión Geográfica-Exploradora and the Comisión Mexicana de Limites con Guatemala (1880–1882), along with other commissions and undertakings. By following these recommendations, the recently chartered Instituto Médico Nacional, it was hoped, would signal a decisive step, not just toward assembling a national flora, but toward putting this information to use for the good of the Mexican people.

A "TRULY NATIONAL MEDICINE"

The IMN's secretary, Secundino Sosa, explained that the goal of the institute was to develop a "truly national medicine" that could "celebrate ownership of truths discovered within, and perhaps applicable only to, this country." Pointing to Mexico's distinctive characteristics and rich variety of natural resources, he asked, "how could we not come to create, though it might require a great deal of difficult research, a national science?"[48] To this end, the IMN published a remarkable five-volume work titled *Datos para la materia médica mexicana* (*Information for Mexico's Materia Medica/Pharmacology*). Issued at intervals between 1894 and 1908, the collection provided extensive analysis of over seventy native plants, accompanied by botanical illustrations.[49]

The first volume, which was the longest and the most well known, was written through the collaboration of two dozen Mexican researchers and attracted international attention when the government presented it at the 1894 Louisiana Purchase Exhibition in St. Louis, Missouri. Similar to the way Velasco's paintings projected the idea of a coherent national nature, the members of the IMN hoped to broadcast to an international and domestic audience the idea of a coherent national medicine. This national medicine would be based on the exploitation of native flora through the merger of indigenous vernacular knowledge and modern scientific practice.

To this end, *Datos para la materia médica mexicana* offered well-researched, effective native alternatives to both expensive imports and to certain folk remedies that the scientists of the IMN had found to be ineffective. Researchers took special interest in identifying treatments to ailments that plagued children and the poor, such as typhus, yellow fever, respiratory and gastrointestinal problems, and even mental health.[50] They hoped to meet the needs of everyday Mexicans who could not afford imported medical treatments, while also fueling nationalist pride in native plants and encouraging internal economic growth. According to its authors, "the many doctors who practice in towns far from the capital will," for the first time, "be able to reliably count on medicines within their reach."[51]

The plant descriptions in the multivolume work followed the model established by *La Naturaleza* (see chapter 6), requiring that every one of the plant monographs address each of the following categories: "Historical, Botanical, histological and medicinal analysis, chemical Analysis, physiological Effect, Therapeutic applications, Mode of administration and dosage, and, finally, Bibliography" (caps from original source). The historical section was especially attentive to any information that Hernández might have gathered about particular plants during the 1570s. It then traced all mentions of each plant from Hernández through the Sessé and Mociño Expedition to the modern era.[52]

The authors insisted that the information their book provided was merely one stage in a larger process rather than its conclusion. This is why they did not arrange the plants in any order other than that in which they completed their study of each species. They also stressed that they tested previously known uses attributed to the plants, confirming some effects and disproving others. And they expressed the hope that later researchers would build upon their detailed investigations to propose and test additional uses for each species.[53]

The botanical illustrations within *Datos para la materia médica mexicana* were created by Adolfo Tenorio (1855–1926). Like his friend and teacher José María Velasco, Tenorio practiced both botanical illustration and landscape painting, though he seems not to have excelled at either at the same level as his mentor. He also devoted himself to figure drawing, photography, and science. He even studied violin and piano in the conservatory. At the IMN, Tenorio drew botanical specimens as well as humans who were suffering particular ailments, and continued as a scientific illustrator after the revolution.[54]

The botanical illustrations that Tenorio produced for the IMN adhere more strictly to Linnaean conventions than did the illustrations produced by the Royal Botanical Expedition a century earlier. This likely is a result of the fact that that expedition's draftsmen, Atanasio Echeverría and Vicente De la Cerda created their images as part of the process of exploration of the botanical frontier (see chapter 4), whereas by Tenorio's time most of these plants had been repeatedly illustrated and there now was broad consensus about their taxonomy and about the primary characteristics that should be emphasized for each. Nonetheless, a small number of Tenorio's images seem to intentionally flout Linnaean norms. For example, his illustration of *Valeriana ceratophylla*, which had been previously illustrated by the Royal Botanical Expedition (item 0684 TCHI), depicts the plant as though it has just been dug up with a shovel. Its slightly mangled stems rise from the plant's root ball—still in a clump of excavated soil—which rests on a grassy field.[55] The reason for such aberrations is unclear, though one can speculate that it might have been done to help rural doctors recognize what the plant looked like in its natural state.

Historian of science Angélica Morales Sarabia and chemist Patricia Aceves Pastrana have found that, even when aided by the IMN's guidebook, rural doctors and average citizens had difficulty distinguishing between similar plants. This hindered the IMN's efforts to convince rural practitioners to adopt its text. One solution would have been for the IMN to be more precise about where and when particular plants could be found in the wild, and to provide instructions about how to avoid confusion with other plants that might be growing in that region during that same season. The IMN also could have looked more deeply into which plants in each region were known by the same vernacular name despite being different species. Unfortunately, the IMN did not have the resources for such ambitious field research.

One proposed solution was for the IMN itself to take on the task of collecting, preparing, and labeling simples and compounds for distribution to rural regions. As an institute primarily focused on research, it lacked the ability to undertake such a project on its own, despite its efforts. In 1907, the IMN attempted to partner with English and Spanish drug companies for such distribution, but the Secretaría de Fomento blocked the plan on the premise that the purpose of the IMN was research, not marketing. The unexpected death of two of the IMN's leaders—Secundino Sosa in 1901 and its founder and director Fernando Altamirano in 1908—further impeded the institute's progress. Then, after the Mexican Revolution, many of the IMN's top researchers left for newer research institutions, such as the Universidad Nacional de México (founded 1910) and the Sociedad Química Mexicana (founded 1926).[56]

While the IMN failed to convince rural people across Mexico to adopt its *Datos para la materia médica mexicana*, the descriptions within its pages are revealing about how members of the IMN viewed their research efforts.[57] Rather than employ a third-person voice that would create a sense of distant, unquestionable scientific authority, the authors demonstrated a notable preference for a first-person eyewitness approach similar to that which had come into favor among scientists recounting their rural expeditions.

They extended this first-person expeditionary voice from the field into the laboratory and hospital. We can see an example in E. Armendáriz's twenty-page monograph about a plant known as *el inguande* or *bocconina*. He writes that, "Dr. Mejía, who I had spoken to about bocconina, invited me to an operation . . . on an impoverished ill patient; I proposed at that time that we use bocconina in place of chloroform, to which he promptly agreed. I applied an injection to the patient as close as possible to the location where we were to cut an incision. . . . The effects of this injection were an initial intense pain that faded little by little until it completely disappeared within five minutes. After that time, the anesthesia was complete and the operation could be done without the patient feeling any pain and without the need for another injection."[58]

This first-person, eyewitness approach was an intentional departure from the kind of descriptions to which members of the Enlightenment-era Royal Botanical Expedition had aspired, and it lent weight to the goal of identifying a specifically Mexican science, rooted in the culture and natural world of the republic. Moreover, it established the author as an intrepid explorer of the natural and scientific wonders not just in his own countryside, but also in Mexico's laboratories and medical research hospitals.

CONCLUSION: ENVISIONING THE NATION

As scientific intellectuals promoted a nationalist science, they searched for ways to organize their findings and tell the story of Mexican nature in a manner that served the needs of the republic. Visual culture played as crucial a role in their process as it had in the Renaissance- and

Enlightenment-era expeditions of discovery. But in place of the dispassionate descriptions preferred by the members of the Royal Botanical Expedition, these nineteenth-century nationalists penned Romantic narratives of exploration and discovery, in which visual description and first-person accounts emphasized the authenticity of their experience and the coherence of the Mexican Republic. In addition to conjuring visual images with words, they also turned to illustrations that invited the reader to join them in bearing witness to the scientific and picturesque natural wonders of Mexico.

Just as scientists turned to art to convey their message, so artists turned to science. This marriage of science, art, and nationalism was exemplified by the landscape paintings of José María Velasco. His canvases gave visual testimony to the wholeness of Mexican nature, producing a nationalist visual vocabulary that would not be equaled until the 1920s, when Diego Rivera drew on an ethnicized view of popular culture to create an aesthetic language that for the first time told people of all backgrounds that, together, they formed the real Mexico.[59]

The IMN carried the naturalists' project from the countryside to the laboratory by attempting to integrate the numerous native simples into a specifically Mexican approach to medicine and pharmacology. Its narratives emphasized the visual process, describing what the researchers witnessed in the course of their investigations, and they offered descriptions and illustrations to help their readers locate and administer the correct simples. Together, these explorers, artists, and medical researchers conjured a world in which plants and other aspects of nature from across the republic fit together to form a coherent whole that was distinctively Mexican.

Despite the nationalist ambitions of these scientists and artists, and even though the IMN sought to minister to the medical needs of the popular classes, their efforts centered the agency of the white and mestizo middle and upper classes. Rather than attempt to listen to the voices of the popular classes, they tried to bring them into line with their top-down nationalist vision, while remaining ambiguous about if and how the masses might contribute to the future shape of the nation. This contrasted with Germany, where the linkage between nation and nature emerged from the provinces and cut across economic classes. As German historian Thomas Lekan explains, "By hiking in the *Heimat*, studying a local bird species, or creating a plaque marking a unique landmark, individuals" from across German society "gave the nation a tangible, emotive quality that linked individual life stories to the collective memory of *Heimat*, enabling provincial Germans to visualize, touch, and even smell their particular region as part of a larger geographical entity: the German nation."[60] In Mexico, by contrast, the linkage of the Romanticized and scientifically analyzed natural landscape, on the one hand, and the republic, on the other, was a top-down process coordinated by middle- and upper-class scientific intellectuals. Advocates often operated with a narrow notion of which people within the Mexican territory were members of this nation, and in whose name they claimed regional natural resources and landscapes.

Their project to merge scientific precision and Romantic expansiveness conveyed an understanding of Mexican nature that invited scientific intellectuals to pay attention to the landscapes around them. As they did so, these intellectuals became increasingly concerned about the rate of environmental degradation that was ever more evident. As the next chapter shows, nowhere was this concern for environmental destruction and the need for conservation more pronounced than in debates about the fate of Mexico's forests and who was to blame for their decline.

CHAPTER 8

SAVING MEXICAN NATURE

In 1873, the military hero turned historian Manuel Balbontín (1824–1894) invited his readers to join him on an imaginary trip across Lake Texcoco as it would have appeared to the first Spanish conquistadors. He asked us to picture the beauty of the lake "surrounded on all sides by magical gardens," crisscrossed by bridges that led to the Aztec capital of Tenochtitlán. But upon opening "your eyes to reality," you would find yourself overtaken by a profound disappointment. In place of a vast picturesque lake, you would see only a "stagnant, muddy puddle," polluted by "the fecal spills of the great city." What happened to cause this? he asks. If the "conquest had never happened, would Texcoco Lake have preserved the utility and beauty that it used to have? Would the Indians," compelled by "political and military interests in its conservation" have saved it from becoming "shamefully littered with the detritus of the capital"?[1] Balbontín and other Mexicans felt themselves called to action, not just by the condition of the central valley's lakes, but even more so by the accelerating devastation of the nation's forests.

In this call to action, they increasingly saw their fight to save Mexico's forests as a struggle for the future of the republic. This chapter addresses the questions of what they felt Mexico's natural world should look like, how they came to understand it as in decline, who they felt was to blame, and what potential solutions they conceived.

DISCOVERING GRANDEUR IN DECLINE

SMGE member Eduardo Ruiz argued in 1872 that "a town should not be . . . depressed because the bowels of its earth are barren of silver or gold," so long as it also possesses "fertile land, forests that produce abundant wood" or "waterfalls and perennial springs that power machines and irrigate fields" and rivers that offer "comfortable and safe means of communication, and . . . easy export of products." These, he argued, are the real sources of "well-being and wealth for the people." Ruiz looked beyond utility, to also appreciate the aesthetics of nature, praising the many "lovely trees . . . that merge to form extensive forests," welcoming us "with the freshness of their shade, with their flavorful fruit, and with the aroma of their flowers."[2]

In his 1897 exploration of central Veracruz, Julio Zárate (1844–1917, politician, lawyer, and historian) moved even further beyond economic utility in his love of nature. In the pages of the SMGE's *Boletín* he praised the beauty of Orizaba Peak, which he preferred to call by its indigenous name Citlaltépetl, whose meaning he wrote is "star mountain." Zárate described how the mountain "rises 5,300 meters above sea level, and its snowy cone emerges immaculately from the waves of the Gulf, announcing to sailors the long-awaited end of their journey thirty or more leagues before they set foot on the shores of Veracruz." He offered a similarly picturesque description of its "companion," Nauhcampatépetl, known in Spanish as Cofre de Perote.[3]

As much as he admired these peaks and the "rich vegetation" that "covers the mountains," he was even more captivated by the "showy and varied" forests that separate the mountains from the shore, where "plant species of the two areas [the temperate mountains and the humid lowlands] become confused and intermingled" to form an "impenetrable forest." The only thing that these "virgin forests" allow to pass through them are the "crystal clear waters of the rivers." It is here, he argued, that "the traveler finds himself emotionally overwhelmed as he is confronted by the magnificence and splendor of nature, vigorous with life," which "evokes ideas of the eternal."[4]

As in other parts of the world, the expansion of the railroad increased interest in the countryside among Mexico's urban middle and upper classes, who could now witness the beauty, as well as the destruction, of the republic's forests. Some traveled on the new railway system to see the country with their own eyes, but, like U.S. and European audiences, most saw their landscape by consuming the growing number of visual images in circulation.[5] One of the most influential sources for such images was the *Álbum del ferrocarril mexicano: Colección de vistas pintadas del natural* (1877) by lithographer Casimiro Castro, who already had earned fame for his richly illustrated book *México y sus alrederores*. Republished many times since it first appeared in 1855, *México y sus alrederores* made images of the people and vistas of Mexico available to the urban reading public, and its fame drew a ready audience to Castro's new collection of lithographs.[6] To accompany Castro's images in *Álbum del ferrocarril mexicano*, the cartographer Antonio García Cubas wrote unabashedly Romantic descriptions of Mexico's landscapes, as well as a history of the development of Mexico's railway system.[7]

Castro's colorful lithographs celebrated the railroad system as well as the spectacular natural and human landscape in which these engineering marvels were situated. Similar to the naturalists' images described in the

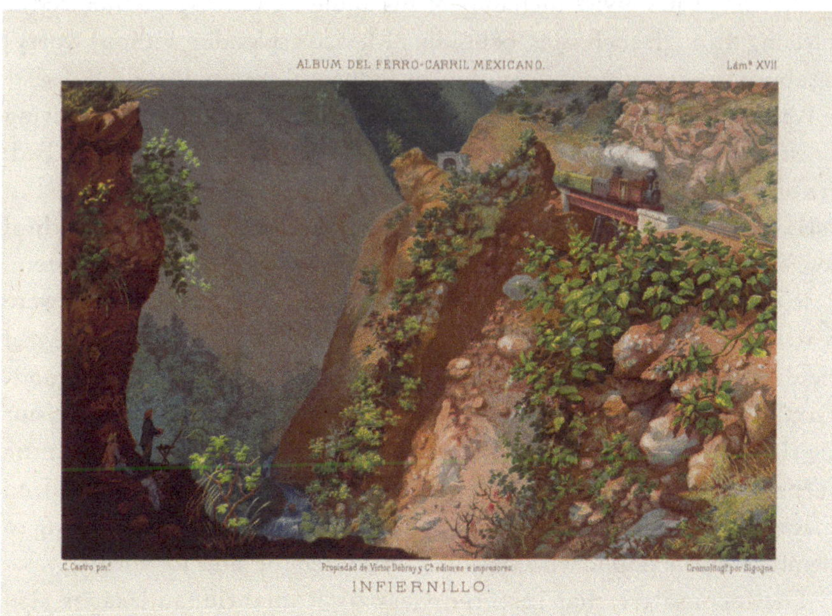

FIGURE 30 Casimiro Castro, *Infiernillo*, in *Álbum del ferrocarril mexicano: Colección de vistas pintadas del natural* (Mexico. 1877), Lam. XVII, Amherst College Archives and Special Collections.

previous chapters, Castro's prints invite the reader to join the explorer as eyewitness. His depiction of Infiernillo (figure 30), for example, celebrates the construction of a rail line through difficult terrain, while highlighting the beauty of the vegetation and landscape. The artist in the lower-left corner of the lithograph prepares to paint the scene. In what seems to be a direct reference to *Kindred Spirits* (1849) by the U.S. painter Asher Brown Durand, Castro's image invites his viewer to imitate the artist's admiration of the aesthetic beauty of the scene. But whereas Durand portrays the painter accompanied by a poet, together beholding U.S. wilderness, Castro shows a painter accompanied by a naturalist attired in the standard expeditionary garb of the era (round-brimmed hat, high-quality sarape, fitted hiking trousers, and leather boots; compare with chapter 7, figures 28 and 29). Together, they admire the engineering marvel and natural world. Like José María Velasco, Castro defines the picturesque not by wild nature, but by the nationalist merger of nature with science and technology.

Castro's prints invite members of his audience to imagine themselves traveling through such sites, beholding Mexico's splendor without worrying about the disruptions unleashed by industrialization. Plate XX, *Puente de Weimer* (figure 31), for instance, shows the modern railway in a thriving forest that is damaged only by the unavoidable cutaways for the rail bed. In contrast to the disruption to tradition that Velasco had depicted in *Cabrio de San Angel,* the man walking the line, accompanied by his faithful dog, suggests harmony between the campesino and the new rail system. Castro's images offer no hints about the suffering of the masses who were bearing the brunt of Porfirian modernization. Neither do they reveal the destruction caused by the frequent forest fires ignited by the wood-burning locomotives, which destroyed vast woodlands along the route, nor the clearcutting that was done to provide the railroad company with the sleepers for the railroad bed on which the countryman walks. Idealized images such as these perhaps made the experience all the more jarring to members of the public who ventured out to see the sites for themselves.

The locomotive that inspired Castro's Romantic landscapes also changed how naturalists moved through rural Mexico. In an article published in the journal of a relatively new natural history organization called Sociedad Científica Antonio Alzate (founded in 1884), Fernando Altamirano recounted an expedition he undertook in 1889 through the

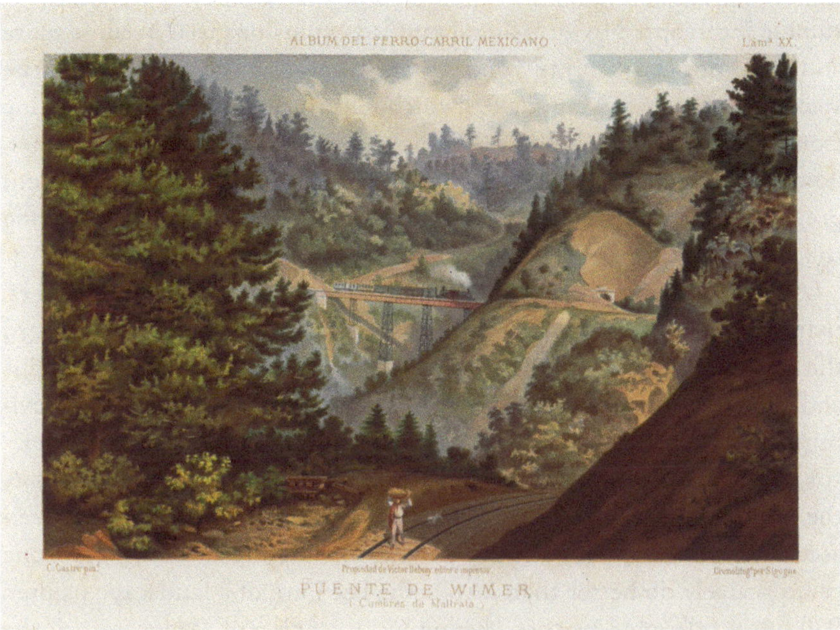

FIGURE 31 Casimiro Castro, *Puente de Weimar*, in *Álbum del ferrocarril mexicano: Colección de vistas pintadas del natural* (Mexico, 1877), Lam. XX, Amherst College Archives and Special Collections.

state of Puebla. He set out by train from Mexico City, accompanied by several other naturalists, including Guillermo Beltrán y Puga and the landscape artist and botanical illustrator Adolfo Tenorio. During the journey through the forested landscape, Altamirano collected information about each stop, including the elevation, local plant species, and the distance from the previous stop. Emphasizing the sense of adventure, he described how a recent storm had destroyed a railroad bridge, forcing him to continue on foot at Necaxa, where among three hundred inhabitants, only twenty spoke Spanish, while the rest were "indios who only know mexicano [Náhuatl]." To continue on the journey, "we had to ford a river" to reach the gorge. Bringing the reader into the scene, he wrote: "Upon advancing [into the gorge] you discover the riverbed that runs 300 meters below the road, forming a canyon that on both sides is covered with exuberant vegetation that gets lost among the mist that always covers such places. Advancing further you start to hear a dull rumble interrupted by occasional muffled explosions that become more and more noticeable, until you arrive at a point where you discover the splendid Salto de la Ventana . . . which falls to the bottom of the ravine from a 219-meter-high cliff."[8] The narrative continues in this manner, flowing smoothly between an emotive tone of intrepid adventure and precise calculations that seem to serve no clear purpose other than giving the account a sheen of scientific precision.

As naturalists reached the far corners of Mexico in growing numbers, thanks to the expanding railroad network, and turned their eyes toward the aesthetic beauty of the forests that covered the republic's mountains and lowlands, it became impossible to ignore the deforestation and environmental degradation. In his survey of the declining condition of Mexico's forests and natural environment, Balbontín quoted Eugenio Landesio, the father of Mexican landscape painting and mentor to José María Velasco. If one is to critique the destruction of a picturesque landscape, one could scarcely do better than invoke the keen eye of a landscape painter. Balbontín informed us that, during his ascension of Popocatépetl, the artist had observed that the oyamel firs that once covered the mountain had disappeared, leaving only ocote pines. But now even these were being eliminated. He described the ground littered by trees that had been cut down and left to rot. And many of those that remain "'appear to be injured at the base by turpentine exploiters.'" Continuing to quote Landesio, Balbontín wrote that soon even these last "'stands of ocote will be destroyed,' and it will take 'centuries before the forest recovers to the vigor that it had just ten, thirty or fifty years ago. This realization caused me sorrow.'"[9]

"LET TREES COVER OUR LAND!"

The accelerating pace of deforestation during the last part of the nineteenth century prompted both the SMGE and the SMHN to declare it to be among their top concerns. The SMGE's *Boletín* led the way with an early call to action in 1869: "The rapid and daily destruction of forests across the country demands the attention not just of men of science," but of each Mexican who can look out his window at mountains that "not long ago were covered with forests," but which today "appear denuded or planted with barley or wheat."[10]

Boletín editors were concerned that not only were the trees disappearing, they were being squandered. "The consumption of firewood by bakeries, bathhouses, locomotives, ceramic kilns, and so forth is enormous and each day cuts down or destroys double or triple the number of trees that would be necessary if they were harvested in an orderly manner." They predicted that this deforestation would lead to scarcity of lumber and firewood, which in turn would leave large numbers of laborers in mines and other industries without work. Soon Mexico would become dependent on the United States and Russia for its lumber, and "only a millionaire will be able to build a home."[11] José Hilarión Romero Gil (1821–1899, lawyer and economist) similarly denounced the wasteful destruction of the forests. Invoking Alexander von Humboldt, he argued that a long history of exploiting wood for mines, churches, palaces, and homes with no effort at conservation had transformed Mexico's great forests into fields and barren wastelands.[12] Like the *Boletín* editors, Romero Gil implored his countrymen to conserve Mexico's remaining forests by exploiting their resources more rationally.

These appeals were not just pragmatic, but also emotional. The editors spoke lovingly of the lakes and mountains of the "Valle de Tenoxtitlan [*sic*]," warning that "the sweet shade and healthy influence of the forest" was being replaced by dry, barren lands. They called on all sectors of society to "contain the harm and stop this calamity."[13]

This romantic sensibility did not prompt a longing for wilderness, as it did in the United States. Instead, representatives of the SMGE called for forests managed with scientific rigor and defended by the policing power of the state. "Today civilized nations" such as France "have established police forces in the forests and mountains that are quite remarkable," resulting in woodlands whose beauty rivals or exceeds those that "nature had taken centuries to create." For Gil, this demonstrated that the proper deployment of "science, experience, and good administration" could result in effective forest conservation. He called on Mexico to join other "civilized nations" by embracing their methods.[14] Alongside the editorial and Gil's essay, the *Boletín* reprinted the text of a forest regulation from Veracruz for the "conservation and increase of forests" that called for creation of two kinds of forest guards—*celadores de bosques* and *policía rural*—as well as local boards known as *juntas conservadoras de arbolados* (woodland conservation councils) to oversee the implementation and enforcement of conservation measures.[15] Antonio García Cubas clarified that, while the SMGE's commitment to the scientific management of forests grew out of economic considerations, it also was inspired by a desire to preserve the beauty of Mexico's natural patrimony.[16]

Ignacio Ramírez, Gumesindo Mendoza, Luis Malanco, and Ignacio Cornejo joined together to proclaim "Let trees cover our land!" A "clamor," they insisted, that was "heard not just in Mexico, but in all civilized nations."[17] They elevated Germany and France as models of conservation. France, whose people, they claimed, had a druid-like veneration for their forests, borrowed insights and experience from Germany until it developed its own effective forest codes, enforced by police. And the United States, according to the authors, was thriving because it honored the sanctity of private property, which encouraged owners to effectively manage their holdings. As for England, it had nothing to teach regarding forestry, since it simply relied on resources extracted from its overseas empire while it destroyed its own woodlands.[18] The tenor of their appeal might lead us to think they might be calling for radical conservation measures. Instead, they dug in their heels in defense of economic liberalism, deregulation, privatization, and *positivismo*.

Driven by their anti-campesino biases, these SMGE members argued that Mexico's problems resulted from communal landholdings by indigenous campesinos, where "communism (*comunismo*)" quashed personal initiative, rational exploitation, and progress. They called for aggressive privatization of communal lands and the removal of state regulations that they claimed infringed on the rights of private landowners. What Mexico needed, they insisted, was scientific planning rooted in positivismo, land privatization, and prioritization of the provision of lumber and fuel to the country's growing industries and railroads.[19] In place of regulations, they called for education of the public about the importance of woodlands, the creation of a national school of forestry following the Franco-German model, and government incentives to encourage private owners to plant trees.[20]

When Ramírez, Mendoza, Malanco, and Cornejo called for reforestation in the name of the Mexican Republic, they did so in a manner that invalidated rural communities' right to the land and denied their long history of efficiently managing the landscape. Their assumption that campesinos were a threat to nature and to the republic was part of the elite process of expropriation. An instructive comparison can be drawn between their presumptions and those of French imperialists in North Africa. When the French came across desert regions in Algeria, Morocco, or Tunisia, they presumed that these places had previously been forested. And they blamed "ignorant peasants" for having destroyed these forests, contributing to the desertification of the landscape. As foresters became concerned about the condition of forests in the Alps and Pyrenees, they extended this same anti-peasant attitude into France itself.[21] In a similar vein, Mexican elites presumed that the natural condition of the Mexican landscape was to be forested. When it lacked forests, they blamed indigenous peasants. Moreover, they aspired not for an untrammeled forest nor for a woodland that could serve the needs of multiple sectors of society, but for protected resources reserved for the exclusive use of the dominant classes.

Elites used their bias to justify not just discursive claims on campesinos' land and resources, but also actual physical displacement and land expropriation. Between the time of Ramírez's 1870 report and 1907, domestic food production to feed the growing population collapsed. State policies intentionally replaced subsistence and domestic food production with lucrative export-oriented agriculture that benefitted Mexican elites and foreign investors. The subsistence farming crisis was accompanied by massive, and notoriously corrupt, transfers of land from local communities to private owners. This process greatly accelerated after Porfirio Díaz affirmed his dictatorial authority in 1884 and began enforcing the 1883 Land Reform Law by granting land survey companies a third of any communal land they chose to survey, with another third sold at auction to Díaz's loyal cronies. To accelerate railroad construction, the government also granted railroad investors free access to any land they needed that either had not yet been transferred from communal to private ownership or else was in tax arrears. By 1910, on the eve of the Mexican Revolution, between 80 and 98 percent of Mexico's rural families had become landless peons laboring under miserable conditions on elite-owned haciendas.[22]

In 1870, Manuel Payno took issue with his colleagues' prescriptions. He undertook his own study of the deforestation problem and of existing conservation legislation. Lamenting that "around the main population centers" throughout Mexico "there no longer is anything to see but nakedness and sadness, especially in the Valley of Mexico, Chalco, Texcoco, Toluca, San Luis, and, generally, all our cities," he encouraged his audience to imagine how lovely the surrounding mountains must have been back when they were densely forested.[23] He took the reader on a detailed tour on foot through the paths and villages in the mountains around Mexico City, observing details regarding the current condition of the forests compared to how they once looked. With this tour, he argued that even though we might not yet have hard data we gain a clear "idea of the barbarous treatment inflicted upon Mexico's wondrous works of nature [*obras maravillosas de la naturaleza*]." With a nod toward the emerging field of ecology, Payno also worried about the impact that deforestation was having on other plant life, including medicinally useful plants that thus far "have escaped the research of botanists," many of which are found nowhere else in the world. He argued that if nothing is done to change the current trajectory, serious scarcity of wood will result, followed by calamity. "Mexico's nature is fertile. No one denies it." But the question, according to Payno, was "Which is stronger, the fertility of nature, or the barbarity and greed of man?"[24]

While Payno agreed with his colleagues' call for the planting of new trees, he insisted that that alone would be inadequate. Contradicting their call for deregulation, he argued that what was needed was new enforceable conservation legislation. From the time of the "conquest through today, not a single tree has been planted in our forests. Our huntsmen and large landowners would take as a lunatic any man who they might see planting trees in the mountains. And, in our forests, no tools are known other than the axe, saw, adze and fire."[25] He discussed the emerging science of tree rings—which we today know as dendrochronology—to figure out how long a tree must grow before it can provide useful lumber. Because a tree takes so long to grow, even if people immediately stopped destroying the forest and started planting trees, it would take many years before those trees could be harvested for lumber. To fill the future needs of the republic, the reproduction of the forests and mitigation of the threats they face had to be better understood, according to Payno, so that policy could be shaped around facts rather than presumptions.

In a review of forestry codes from the 1500s through his own day, he found every one of them too vague and unrealistic to have been effective. He gave numerous examples of laws that imposed all kinds of restrictions on indigenous people cutting trees for firewood near mines, but ignored the fact that "it is not the Indians, but the English and wealthy Mexicans who chop down the forests" for use in mining, industry, church construction, and railroads.[26] Payno urged scientists and government offices to collect data on the consumption of wood and the sources of that consumption—he provided a detailed list of the exact kinds of data he sought—so that the problems could be better understood and effective management solutions at last be formulated.

It is instructive to note that even within the SMGE, as demonstrated by the disagreements between Payno and Ramírez, intellectuals did not speak with a single mind. Despite broad agreement about the horrors of deforestation, the benefits of land privatization, and the power of positivismo, Payno strongly disagreed with his colleagues regarding the desirability of regulations, the role of indigenous people, the future of the commons in forest conservation, and whether education and tree planting, on their own, were adequate solutions to the problem.

In 1873, the SMGE officially listed reforestation as one of its highest priorities. Manuel Balbontín, the Liberal war hero who fought against the U.S. invasion and the French Empire, forcefully denounced the "crude war" by "the inhabitants of the Republic" against the forests without consideration for the "grave harm against the Nation from such barbarism" or the "awful inheritance we leave to our descendants."[27] He extensively quoted observers from across Mexico to show the widespread destruction of "the works of nature" just so peasants could "plant a milpa" on land that a year later will lose its fertility and become a "worthless wasteland." This had resulted in "harm to mining, natural springs, health, and industry."[28]

Balbontín lamented not just the loss of trees, but also the destruction of what today we might term an ecosystem. We see this, for example, in his discussion of the Oasis de Aguanueva. He recounted how, during the U.S.–Mexican War, his troops found respite in this oasis after the stalemated battle of 1847. It offered them "a place to quench their thirst and protect themselves from the heat of the sun during the four days they remained camped there." Balbontín asked the reader to imagine his surprise when he returned to this same spot twenty years later and discovered that the oasis had disappeared. "Where was it? What could have happened to it? Sirs, shockingly, the owner of the hacienda had ordered that [the trees] be chopped down!" The disappearance of the trees did away with the oasis, and now, in place of the "stream of pure water that saturated the air with pleasant freshness, the inhabitants" had to dig a well "so as not to die of thirst."[29] The German-born naturalist and SMGE member Hugo Finck similarly described denuded landscapes around Orizaba, Veracruz, that he claimed had been covered with oaks and "colossal" conifers just a few decades earlier. By 1874, "everything has been destroyed, and at this rate, within a few years our mountains' most beautiful adornment will be gone."[30]

Balbontín presented this heedless destruction of the forests as an affront to nature, Mexican history, the republic's ability to defend itself against foreign invasion, his own personal memories, and the people who continued to live on that land. Examples of such destruction surround us, he argued. Even in Chapultepec, stands of sabinos (a kind of cypress) had been burned to the ground. Across the country a "merciless war is waged against trees. Their bark is stripped off, their feet are drowned in filthy water, then they" are ripped out of the ground by their roots. And every year Mexico City was spending huge sums planting new trees, and yet except in the Alameda and Zócalo where there were protected by guards the parks lacked trees as a result of the way people mistreated and destroyed them. Balbontín argued that such crimes should be punished, but that the more effective approach was education to "persuade the people about the great importance of woodlands" for the economy, public health, and general well-being. He argued that his fellow Mexicans needed to learn "the great advantages to be gained by doing the exact opposite of what has been done up until now regarding forests."[31]

"THE FUNDAMENTAL CAUSE OF THE DESTRUCTION OF THE FORESTS": TOWARD A RADICAL POSITION

The SMHN soon joined the SMGE in this effort by adding forest conservation to its own mission statement in 1890. But the SMHN's journal *La Naturaleza* expressed skepticism regarding the SMGE's free-market approach. Instead, it leaned toward a comparatively positive view of the needs and activities of indigenous people and campesinos, and took an overt interest in the new field of "ecology," which it also added to its mission statement.[32]

The term "ecology" was coined by German naturalist Ernst Haeckel in 1866, drawing largely on ideas that Alexander von Humboldt had formulated during his travels in Mexico and South America. While the first book establishing the foundations of ecology as a field would not be published until 1895, by the Dutch botanist Johannes Warming, the SMHN already was eagerly moved forward in its defense of both forests and ecosystems.[33]

During the early 1880s, even before officially incorporating these concerns into its mission statement, the SMHN had already created a special commission to develop a plan for the reforestation of the valley of Mexico and the restoration of local ecologies. The commission was headed by Mariano Bárcena and comprised Miguel Pérez (subdirector del Observatorio Meteorológico Central), Manuel Urbina (1841–1906, physician), José Ramírez, and José Segura (1846–1901, agronomist). After completing its study of the various climates and soils within the valley in 1884, the commission submitted a list of trees appropriate for each terrain. With the exception of eucalyptus, all the trees on its list were native to Mexico—though, as the authors pointed out, not necessarily to the central valley. Members of the commission explained that the reason they made an exception for eucalyptus was that it quickly adapts and spreads, thus serving as a safeguard against "such moments when our forests disappear with alarming speed."[34] So severe was the pace and intensity of deforestation, along with the negative effects that come with it (such as flooding and changes in hydrology), that a quarter of the trees recommended by the commission were varieties of eucalyptus suited to these exigent circumstances.

After it revised its mission statement, the SMHN linked forest conservation to broader ecological concerns. In 1895, for example, Manuel Villada delivered a speech arguing for the need to conserve forests to protect the supply of surface water needed to moderate local climatic conditions and for public health.[35] In recognition of the organization's expertise, the government began commissioning SMHN members to conduct special studies into ecological matters. One such study, commissioned by the Secretaría de Fomento, was to figure out why the ancient trees of Chapultepec were dying. This committee, headed by Mariano Bárcena, also included Miguel Pérez and a certain Sr. Esparza (conserje de Chapultepec). The committee concluded that the problem was caused by a drop in the water table because of overexploitation—not by the introduction of eucalyptus trees, as some had been claiming. To mitigate the problem, the committee recommended interspersing saplings among the ancient trees to preserve moisture and moderate the local climate.[36]

As the impacts of deforestation became more evident, some SMHN members pulled no punches. In 1895, Fernando Altamirano, perhaps emboldened by the accolades the government had bestowed on him for his achievement at the Paris world's fair of 1889 (see chapter 7), pointed his finger at people who did not take kindly to being called out. His report began with a rallying call for conservation. "There is no art that does not require wood; there is no industry nor locomotion without fire. Hygiene, meteorology, and hydrology show us daily the need for forests; in a word, without them nations would perish." Warning that "Mexico will perish because of its destruction of its forests!" he called on his countrymen, starting with his fellow members of the Sociedad Mexicana de Historia Natural, to commit themselves to reforestation.[37]

The worst decline, Altamirano argued, had been during the last twenty years, within his audience members' living memory. To put this recent decline into context, he reconstructed a longer historical perspective, starting with the Aztec era, when the hills of the valley were covered with forests. Citing Hernández, Altamirano noted that the ancient Nahua had special affection for the forest, as well as for "introducing new species brought from distant regions" of Mexico.[38] And Antonio Alzate, in the late 1700s, "already was lamenting in his *Gacetas de literatura* the immoderate cutting down of trees." Deforestation, according to Altamirano, started in the plains, then extended into the hills, and "now it is reaching the mountain tops," the final refuge for Mexico's once grand forests. "Today the vast pine and oak forests have almost disappeared, along with the majestic cedars and oyameles [a type of fir] that were the jewels of our mountains and the source of life for many communities. Owners of woodlots no longer even have anything left to sell," and the forests that previously covered the "hillsides have been reduced to a narrow and thread-bare ribbon atop the mountain range, like the last vestige of its old regalia."[39] Altamirano acknowledged many contributors to Mexico's deforestation, including forest fires, diseases and infestations that attack particular species, conversion of woodlands into agricultural land by both rich landowners and campesinos, and climate alterations brought about by the removal of surrounding forests that negatively affected the remaining tree stands, all of which have long worked in concert against the forests.

But the recent acceleration of destruction, he argued, was to be blamed entirely on the railroad companies, who used the trees for railroad ties and "rely on firewood as the only combustible." In a remarkable departure from the free-market, anti-regulatory, and anti-campesino attitudes of most Porfirian-era intellectuals and government officials, including most members of the SMGE, Altamirano, speaking on behalf of the SMHN, argued that this destruction resulted from a lack of regulation and the "complete liberty that has been given to the railroad companies to consume firewood, sleepers, posts, etc., with no tax [*gravamen*] nor restriction of any kind."

The massive and unregulated consumption of wood by industry and railroad companies, he concluded, "is the fundamental cause of the destruction of the forests; it is the true and only [cause] and it demands all our attention" if we hope to resist the relentless "attack upon public resources." As the railroads and industries reaped their short-term benefits, he argued, those who suffered were the forests, small industries, and poor families. Any solution "requires not just forest regulation, but rather a great many other complicated regulations that are difficult to resolve in our current circumstances."[40] With this stance, Altamirano staked out a position for the SMHN that was more radical than anything emanating from the SMGE. This likely alienated the Porfirian *científicos* who controlled government purse strings and perhaps contributed to the eventual decline of the SMHN compared to the SMGE and a new rising organization called the Sociedad Científica Antonio Alzate (which over the course of the next century would evolve into the Academia Mexicana de Ciencias [Mexican Academy of Sciences]).

CONCLUSION: THE END OF THE SMHN AND THE FUTURE OF FOREST CONSERVATION

Though the SMHN's forest conservation recommendations won a growing number of enemies among the country's economic elite, members of the organization in late 1910 seemed on the verge of gaining durable state support to help pursue some of their other objectives. Just two months before the outbreak of the Mexican Revolution, one of its members Jesús Sánchez (1842–1911, doctor and naturalist) announced the creation of the Museo Nacional de Historia Natural, of which he was to be the inaugural director. The new museum, with a generous annual budget, would even house a botanical garden, albeit not on the scale that the SMHN had been lobbying for.[41] The museum was to operate under the auspices of the Comisión Exploradora de la Flora y Fauna Nacionales, a government office that had been created in 1879, but which only in 1908, under the direction of the naturalist Fernando Ferrari Pérez (1857–1933) and the prominent SMHN member Manuel Villada, emerged as an independent department within the Secretaría de Fomento, dedicated to fulfilling the SMHN's goal of at last cataloging all Mexico's native flora and fauna.[42]

Even before the disruptions caused by the Mexican Revolution, the study of natural history—including botany, forestry, and ecology—was threatened by the deaths of many of the SMHN's leading members, combined with a failure to train sufficient numbers of the new generation. The loss of leaders came so quickly one after the other that Manuel Villada wondered if the SMHN could survive.[43] But the more immediate challenge was the revolutionary upheaval that led to the dissolution of the Comisión Exploradora de la Flora y Fauna Nacionales and its museum. Their collections, in an eerie echo of the fate that had befallen the work of Hernández and that of the Royal Botanical Expedition, became scattered across other departments and lost.[44] In 1914, even the SMHN dissolved, along with its nationalist commitment to ecology, conservation, and native botanical patrimony.

Eventually new scientific institutions would emerge, as would new academic departments within the Universidad Nacional de México.[45] Not until after the revolution, during the 1930s under President Lázaro Cárdenas, would the country's scientific intellectuals again join forces in pursuit of conservation of nature as nationalist patrimony, this time with the added mission of social justice inspired by the popular revolution.[46] Nationalist botany attracted renewed interest when scientists of the 1930s began rummaging through the remnants of the Royal Botanical Expedition's herbarium in search of native resources to exploit for the good of the public.[47] The Cardenista state even compiled information about native plants, including medicinals, in its *Enciclopedia nacional popular* (*National Popular Encyclopedia*), originally published in 1935 for distribution to the popular classes across the country.[48] Only later would Mexico return to the question of whether it needed to take action specifically in defense of native species, defending them not just for their symbolic importance, but also for their economic value, especially for the growing global pharmaceutical industry.

Forest conservation faced setbacks but found new advocates, most notably Miguel Ángel de Quevedo (1862–1946), who would become known as "The Apostle of the Tree" in recognition of his commitment to reforestation. Quevedo was born in Guadalajara, but after the loss of his parents the teenager was taken in by relatives in France. After completing secondary school, he studied engineering and forestry at several French institutions of higher learning. In 1887, he returned to Mexico to work on the *desagüe* project, which exposed him to the rapid deforestation of the mountains surrounding Mexico City. After a serious injury during a tunnel inspection, he left the desagüe project to work for a railroad company, which brought him into even closer contact with the hydrological and social harms caused by deforestation. In 1901, Quevedo began his trajectory as a public advocate for reforestation and conservation parks based on French models, a career that would make him a central figure in postrevolutionary conservation.[49] Historian Joachim Radkau has contrasted the centralized French model to the decentralized German model, arguing that, "in the long run, the French centralist system was not well adapted to forest problems, which often required immediate response to poachers, decisions about local species, and the adjudication of petty squabbles between different users."[50] Quevedo nonetheless continued to emphasize the top-down approach of the French model of forestry, which he endeavored to adapt to the new circumstances unleashed by the popular revolution of 1910–1921.

In these debates of the 1930s, conservationists and nationalists would continue to draw on the discourses and legacies created by their sixteenth-, eighteenth-, and nineteenth-century predecessors regarding the intimate connection between indigenous history, native plants, and Mexican science. But even as they drew on long-standing discourses, they operated within a new social terrain. The popular revolution had opened new possibilities for the popular classes to make their voices heard within debates about the destruction of the natural landscape, and how and for whom conservation should be undertaken.

CONCLUSION

Science, Visuality, and Indigeneity in the Rooting of Mexican Nature

According to intellectual and political leaders of the 1920s and 1930s, the depth of the conflict during the popular revolution (1910–1921) had revealed Mexico's failure to forge its people into a real nation. They insisted that to progress as a country and to stave off a replay of the recent violence Mexico now needed a form of modernization tailored to its distinctive cultural and political reality. With this goal in mind, postrevolutionary intellectuals set out to learn about the country's diverse people, above all the indigenous groups, and use this knowledge to help integrate the population into a national community.[1] Yet, as this book has shown, these postrevolutionary intellectuals were not the first to aspire to endow Mexico with a distinctive identity. Nor were they the first to place indigeneity at the heart of this quest.

An environmental historical approach to scientific intellectuals' attempts to study, describe, depict, and conserve Mexico's natural resources reveals that, before the postrevolutionary project that created the country's modern nation-state, people had for a long time already eagerly debated the importance of indigeneity, first for the empire then for the republic. Scientific intellectuals used these claims regarding indigeneity to root themselves and the state within the land of Mexico and to claim dominion over the plants that sprang from its soil while increasingly marginalizing living indigenous peoples.

In the sixteenth century, Francisco Hernández had been transparent regarding how he drew on indigenous knowledge about the natural world. In the eighteenth century, members of the Royal Botanical Expedition elevated the expertise of Enlightenment men of science such as themselves and subordinated indigenous people, even erasing evidence of the expedition's dependence on natives' knowledge. In the nineteenth century, nationalist intellectuals reinforced the scientific practices of the Royal Botanical Expedition while invoking Hernández to recover the indigenous roots that Enlightenment naturalists had obscured. But rather than center living indigenous people, nineteenth-century scientific intellectuals centered themselves and the Mexican Republic as heirs to the land and to indigenous knowledge about nature.

Visuality was central to each of these endeavors. Hernández and his Nahua collaborators not only developed taxonomies that Hernández hoped would be adopted universally, but they also improvised visual vocabularies to embed native and local knowledge within botanical images and in the landscape. The members of the Royal Botanical Expedition did not support this kind of open embrace of either indigeneity or local perspectives, but, as these chapters have shown, they did rely on such native knowledge more than they cared to reveal. And though part of the original charge of their expedition was to recreate Hernández's lost images, they instead simplified the natural world by viewing it and displaying it through what I have referred to as a Linnaean optic. Nineteenth-century naturalists borrowed from both previous endeavors. But rather than acknowledge the incongruities between Hernández's undertaking and that of the Royal Botanical Expedition, they cast both as part of a single, unified lineage of Mexican natural history.

These scientific intellectuals drew on their natural world to help promote a Mexican identity, but they were not engaged in what we now think of as nation formation; that is, they were not attempting to integrate the culturally, linguistically, and economically diverse masses into a popular-based shared national culture. Instead, they used visual descriptions of plants, forests, and natural features to claim Mexican nature and indigeneity in the name of the republic, while justifying the dispossession of the popular classes. In this way, they presented the natural world and the indigenous past as central to Mexican identity, with themselves and other elites as the claimants to this past and captains of its future.

Just as the Royal Botanical Expedition unwittingly created the foundations for Mexicans to imagine an alternative to the imperial structure that the expedition sought to impose, so, too, did the scientific intellectuals of the late nineteenth century unintentionally construct discourses that later would be used by postrevolutionary nationalists eager to imagine a more inclusive alternative. After the Mexican Revolution, intellectuals drew upon these sixteenth-, eighteenth-, and nineteenth-century discourses, even as they claimed to reject the thinking of their colonial and Porfirian predecessors. The internal tensions that characterized each of these projects rendered them into a usable past for postrevolutionary leaders. And these tensions remain fundamental to Mexico's relationship to its natural world as well as to its indigenous heritage.

POSTSCRIPT

Flags, Zones, and Native Seeds

Today, the symbol of ancient Tenochtitlán—an eagle holding a serpent while perched on a nopal cactus—fills the center of the Mexican flag (see figure 32). The story of how this Aztec image became the symbol of the nation, and how its appearance changed over time, is part of the narrative of Mexico's evolving connection to its natural world and to indigeneity. And the same legacy that underlies changes to the flag also extends to initiatives that range from national seed banks for genetic conservation to Mexico's underappreciated role in shaping one of the most important international models for environmental protection.

After the conquest, the Spanish Crown expressed ambivalence about the Aztec's symbol for Tenochtitlán until the end of the eighteenth century, when the Academia de San Carlos de Arte, founded as part of the same effort that led to the Royal Botanical Expedition (see chapters 3 and 4) created an escutcheon that incorporated a version of it, rendered in a Europeanized style and surrounded by swags and imperial symbols. Once the Academia de San Carlos had given its imprimatur, the eagle, serpent, and nopal was eagerly adopted within institutional insignia across New Spain, not just in Mexico City.

The mestizo insurgent leader José María Morelos applied the eagle and nopal (but not the serpent) to a flag around 1813. According to historian Enrique Florescano, who recounts the history of the flag from the conquest through independence, the rebels used the old symbol for Tenochtitlán to signal their desire to reclaim stolen sovereignty, not in behalf of the indigenous population, but for the multiracial factions that positioned themselves as inheritors of the Aztec right to self-rule.

The conservative general Agustín de Iturbide, who declared himself emperor of the short-lived Mexican Empire (1821–1823), initially tried to eliminate the rebel's use of the indigenous eagle and nopal. But he soon accepted a compromise in which the new government would keep the symbol so long as the eagle wore a European-style crown. After Iturbide was deposed and Mexico became a republic, the crown was removed and the serpent returned to the eagle's clutch, inspired by the old Aztec symbols, but now with a new meaning: the triumph of the republic over monarchical tyranny.[1]

As the flag continued to evolve during the nineteenth century, it recovered more of Mexico's indigenous history and native perceptions of nature. During the reign of Emperor Maximilian (1864–1867), the flag situated the nopal within an aquatic environment, as a nod to the capital's location surrounded by Lake Texcoco. After Juárez restored the republic, the government eliminated this change to the flag as part of its purge of Maximilian's interregnum. In the 1880s, when natural history societies were reaching the height of their influence (see chapters 6–8), the government of Porfirio Díaz restored the lacustrine context surrounding the nopal.

In 1916, while the revolution was in full swing and artists were beginning to define a distinctively Mexican aesthetic in art, the eagle and the serpent incorporated an Aztec aesthetic for the first time since the preconquest era, likely modelled after ancient stone carvings being unearthed at that time by archaeologists. In 1934, the flag again was altered. This time the rocky hill was replaced by a Nahua stone glyph similar to the depiction of Mexico-Tenochtitlán in the 1564 *Codex Osuna* (see figure 4 in chapter 2), but bearing the national colors of green, white, and red.

The most recent change was in 1968, at a time when Mexico showcased its modernity when it hosted the Summer Olympics while cracking down on internal dissent and reviving *indigenista* populism.[2] The national colors were removed from the stone glyph to further emphasize the indigenous elements of the symbol. And the blue haze of water used by Maximilian and Díaz was replaced by a Nahua water glyph (see chapter 2), thus bringing the flag into conformity with what Janice Neri has termed "species

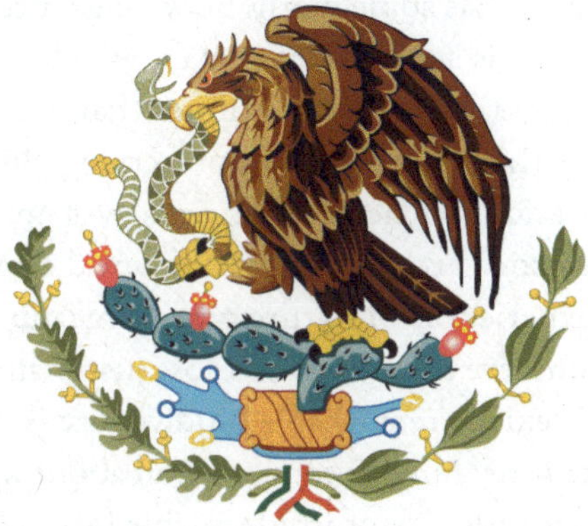

FIGURE 32 Emblem from center of the Mexican flag.

logic" (see chapter 4). This 1968 version of the flag brought nationalist views of Mexican nature and indigeneity full circle by recovering the native modes of representation that Francisco Hernández and the Nahua draftsmen of the sixteenth century had incorporated into their botanical images (but which were rejected by European Renaissance scientists). This evolution of symbolism affirmed Mexico's nationalist narrative of a natural world dense with deeply rooted historical and cultural associations.

The evolution of the Mexican flag reflects Mexicans' changing relationship to nature. Despite these changes, Mexicans, like members of most modern nations, often imagine their connection to their natural environment as eternally inseparable from the timeless nation itself. The United States' association with frontier wilderness has long been the most obvious example of the link between nation and nature, and of the labor required to sustain the mythology. Maintenance of the nationalist illusion of U.S. wilderness requires a lot of work, including on the part of the National Park Service (NPS). In Yellowstone, for example, the NPS is responsible for not just protecting forests and animals but also making it easy for visitors to commune with nature staged as wilderness, and to do so from the comfort of their cars. It is with good reason that Ronald Utt, a senior research fellow at the conservative Heritage Foundation, in 2005 ridiculed the NPS as "the world's largest lawncare and janitorial service." On the basis of this criticism, the Heritage Foundation called for national parks to be defunded and sold off to private investors.[3] While Utt's description of the labor of the NPS has some merit, he wrongly presumes that nature is nature only if it truly is separate from man, which is a false dichotomy, even for U.S. wilderness. He further assumes that, if nature cannot meet this impossible ideal, then it is not worth preserving. But he is looking at it wrong. The real questions we need to ask ourselves not just in the United States but also in Mexico are: What kind of world do we want to live in, and what kind of experiences of nature give us meaning and inspiration? And how can we conserve nature in the most just manner?

Destruction of Mexico's remarkable biodiversity is proceeding at such a pace that the lack of meaningful conversation about what kind of nature Mexicans want creates the risk of irredeemable loss. What is being lost is not untouched wilderness, but human-shaped biodiversity and multiuse environments. For millennia before the arrival of Europeans, people in Mexico distributed seeds and plants along their migration routes and otherwise interacted with the region's fauna and flora, thus creating a level of biodiversity beyond what might have occurred naturally.[4] Rather than harness this biodiversity, we have destroyed it to such a degree that, even if we were to leave the forest to regenerate, it likely would not regenerate with the same level of biodiversity that previously made it so useful to the people who had been living within it. Destruction of the forest, therefore, is perhaps even more lamentable precisely because the forest is not wilderness but the result of complex and long-term human cultivation. There are some promising signs that deforestation in sections of southern Mexico has slowed thanks to grassroots communal management.[5] If this recovery continues, it might reveal whether the forests' biodiversity also returns, or whether a different kind of cover will result.

By the early 1960s, Mexican conservationist Enrique Beltrán (1903–1994) recognized the need for environmental preservation efforts to better accommodate human–nature interactions. This inspired him at the First World Conference on National Parks held in Seattle in 1962 to propose his "zoning system" as an alternative to the national park system for fragile ecosystems. Beltrán, often known as one of Mexico's first modern conservationists, had been mentored by the biologist Alfonso Luis Herrera, whose father Alfonso Herrera Fernández was one of the leaders of the SMHN.

In the "zoning system" proposed by Beltrán, a core would be left undisturbed and accessible only to scientists, while surrounding bands would have incremental increases in the level of exploitation and mixed use. A benefit of this over the park system is that it does not try to erase the human presence from the landscape, nor does it make human use the sole justification for land protection. Moreover, it attracts local support by allowing for a range of negotiated practices in the mixed-use zones. Over the next decade, Beltrán's idea blossomed into UNESCO's Man and the Biosphere Program (MAB) for nature conservation, which was strengthened in 1995 through the Seville Strategy and today is the dominant system for environmental management around the world, with more than forty biosphere reserves in Mexico, twenty-eight in the United States, and over seven hundred worldwide.[6] The MAB offers an alternative model to the much-criticized U.S. view of nature as wilderness isolated from man and the German ultranationalist linkage between landscape and chauvinistic nationalist destiny. Despite the MAB's debt to Mexican intellectual traditions regarding nature conservation, its roots tend to be overlooked internationally and even within Mexico.

The legacy that underlies the national flag and the MAB also has shaped seed conservation programs within Mexico. During the 1930s, the populist government of President Lázaro Cárdenas (1895–1970, president 1934–1940) funded an initiative to help rural farmers by teaching them improved farming practices and providing access to native and hybrid corn varieties suited to each location. Intervention by the Rockefeller Foundation in the 1940s, as part of the so-called Green Revolution, replaced the earlier populist initiative with the market-oriented Centro Internacional de Mejoramiento de Maíz y Trigo (CIMMYT, International Maize and Wheat Improvement Center), a venture dedicated to seed hybridization and genetic modification. More recently, in 2024, partly in response to challenges posed by climate change, a new seed bank called La Casa de Semillas Toxinachcal was founded in Xochimilco. In a partial return to research and ideals drawn from Hernández all the way through the SMHN, this new seed bank focuses specifically on the preservation of thousands of varieties of native plants ranging from corn to nopal.[7] Further momentum has been added by President Claudia Sheinbaum (born 1962, inaugurated October 2024), who declared the nationalist imperative to "protect Mexico's biodiversity." Because maize holds particular importance within this biodiversity and is "part of the national identity," its cultivation must be based on "agroecological management" and "free of genetic modification." To this end, President Sheinbaum endorsed a constitutional ban on genetically modified corn within Mexico, going so far as to assert that "without corn there is no country."[8]

As the chapters of this book suggest, recognition of Mexico as a hot spot of biodiversity has much deeper historical roots than is typically acknowledged. The term "biodiversity" was not coined until 1985, yet recognition of Mexico as a land of unusual botanical diversity is rooted in the sixteenth century, when this notion inspired the Spanish Crown to send Francisco Hernández to New Spain to scientifically document this diversity. It also guided the endeavors of the Royal Botanical Expedition in the eighteenth century. And in the late nineteenth century through the early twentieth, scientists were eager to catalog this diversity before it was destroyed by deforestation and commercial agriculture, and to harness it for the pride and economic benefit of the Mexican Republic. Today's biologists continue to lament the lack of a comprehensive inventory of all the species found within Mexico's borders. And so, the aspiration to create a complete flora of Mexico, which animated botanists of the eighteenth and nineteenth centuries, persists.[9]

In the U.S., France, Germany, Australia, New Zealand, and elsewhere, the connection between nation and nature is often used as a guide for policy, whether in the promotion of wilderness, the idealization of the peasant-pastoral landscape, or promotion of the managed forest. But in Mexico, despite initiatives such as the constitutional change supported by President Sheinbaum, there is too often insufficient conversation among policymakers about the historical relationship between the natural world and the people of Mexico, much less how the dynamics of this relationship should inform conservation policies. Nor do policymakers sufficiently address the question of what kinds of experiences Mexican citizens would like to enjoy with their natural environment. The result is that, despite Mexico's contribution to international models of environmental stewardship, within the country nature conservation frequently is treated as merely a local issue limited to regional conflicts rather than part of a national conversation.[10] By recovering the links between nature and broader discourses regarding Mexican identity, I hope the findings presented in this book can help contextualize these local struggles as part of the national project and as issues that should be of concern to all Mexicans.

NOTES

Introduction

1. Ramamoorthy, et al., "Introduction."
2. On the impact of the wall on mountain lions and other animals on the border, see "Mountain Lion Stuck," and Dell'Amore, "Animals Stopped in Their Tracks." On differing U.S. and Mexican policies toward whales (in which Mexico developed more successful protection programs than did the U.S.), see Dedina, *Saving the Gray Whale.*
3. Benedict Anderson long ago showed that nations are imagined communities constructed to give the illusion of having a deep history. Historian Eugen Weber found that even France, thought to be one of the oldest nations, became integrated into what we might consider a modern nation (that is, one that includes the masses who live within its boundaries, rather than just the middle and upper classes) only in the late nineteenth century. Though Anderson's and Weber's arguments have been debated and refined over the years, their central theses remain intact. Anderson, *Imagined Communities;* Weber, *Peasants into Frenchmen;* Connor, "Timelessness of Nations"; and Guy, *When Champagne Became French.* Regarding how these ideas apply to twentieth-century Mexico, see López, *Crafting Mexico.*
4. Imort, "The Sylvan People," 55.
5. Lekan, *Imagining the Nation in Nature;* Blackbourn, *The Conquest of Nature;* and Schama, *Landscape and Memory,* 6, 82.
6. Thomas, *Man and the Natural World,* 258–59; Mukerji, *Territorial Ambitions;* and Borsay with Sweet, *The Invention of the English Landscape.*
7. Mukerji, *Territorial Ambitions,* 25; Guy, *When Champagne Became French;* and Whited, *Forests and Peasant Politics.*

8. Nash, *Wilderness and the American Mind*; Cronon, "The Trouble with Wilderness"; Miller, *Nature's Nation*; Driscoll, *All That is Glorious*, 19; Steinberg, *Down to Earth*; and Turner, "The Significance of the Frontier," 31–60.
9. Holguin, *Creating Spaniards*; and Casado de Otaola, *Naturaleza patria*.
10. On the development of Costa Rica's identity as the "Green Republic" and current trends that threaten to destabilize the country's democracy, environmental protection, and autonomy, see Vivanco, *Green Encounters*.
11. On Australia, Canada, and New Zealand, see Dunlap, *Nature and the English Diaspora*. This connection emerged decades later in most places as compared to England, France, Germany, Mexico, and the United States. See, for example, Foster, *Washed with Sun*; Thomas, *Reconfiguring Modernity*; and Miller, Thomas, and Walker, eds. *Japan at Nature's Edge*.
12. Carey, "Latin American Environmental History"; and Leal, *Landscapes of Freedom*.
13. Wakild, "Environmental Justice, Environmentalism, and Environmental History"; and Soluri, Leal, and Pádua, "Finding the 'Latin American.'"
14. Brailovsky and Foguelman, *Memoria verde*; Castro Herrera, *Los trabajos de ajuste y combate*; Miller, *An Environmental History of Latin America*; Ortiz Monasterio et al., *Tierra profanada*. Simonian, *Defending the Land of the Jaguar*; Vitale, *Hacia una historia*; Melville, *A Plague of Sheep*; and Miller, "Stilt Root Subsistence."
15. Marquardt, "'Green Havoc'"; Melillo, *Strangers on Familiar Soil*; Soluri, *Banana Cultures*; Tucker, *Insatiable Appetite*; Gallini, *Una historia ambiental*; Carey, *In the Shadow of Melting Glaciers*; Schwartz, *Sea of Storms*; and Walker, *Shaky Colonialism*. Of the second category, see Boyer, *Political Landscapes*; Boyer, ed., *A Land Between Waters*; Dean, *With Broadax and Firebrand*; Edwards, "The Human Impact on the Forest"; Funes Monzote, *De bosque a sabana*; Klooster, "Campesinos and Mexican Forest Policy"; Klubock, *La Frontera*; Pasquel, *Estampas de la Ciudad de México*; Vivanco, *Green Encounters*; Freitas, *Nationalizing Nature*; and Wakild, *Revolutionary Parks*.
16. Arguilez, "Making the Desert Place"; Garavaglia, "La pampa como ecosistema"; Hämäläinen, "The Politics of Grass"; Radding, "The Children of Mayahuel"; Sedrez, "The 'Bay of All Beauties'"; Ezcurra, *De las chinampas a la megalópolis*; and Simon, *Endangered Mexico*.
17. Altieri and Toledo, "The Agroecological Revolution"; Aboites, "Relación sociedad-naturaleza"; Fabricant and Hicks, "Bolivia's Next Water War"; Tortolero Villaseñor, ed., *Tierra, agua y bosques*; Wolfe, *Watering the Revolution*; Wright, "Downslope and North"; Chomsky and Striffler, "Empire, Labor, and Environment"; Santiago, *The Ecology of Oil*; and Studnicki-Gizbert, *The Three Deaths of Cerro de San Pedro*.
18. Carrera, *Traveling from New Spain to Mexico*; Craib, *Cartographic Mexico*; Méndez Reyes, "Instituciones, obras y viajeros"; and Leal, *Landscapes of Freedom*.
19. Impressive works that move in this direction include Bleichmar, *Visible Empire*; and Stepan, *Picturing Tropical Nature*.
20. These include Álvarez Peláez, *La conquista de la naturaleza*; Candiani, *Dreaming of Dry Land*; Cañizares-Esguerra, *Nature, Empire, and Nation*; McCook, *States of Nature*; and Cooley, *The Perfection of Nature*.

21. Collinson, ed., *Green Guerillas*; Downing, et al., eds., *Development or Destruction*; Grandia, *Enclosed*; Martínez-Altieri, "Ecology and the Poor"; O'Brien, *Sacrificing the Forest*; and Revkin, *The Burning Season*.
22. Baver and Deutsch Lynch, eds., *Beyond Sun and Sand*; and Carruthers, ed. *Environmental Justice in Latin America*.
23. Melville, *A Plague of Sheep*.
24. González, "Discurso sobre el Estudio de la Botánica."

Chapter 1

1. Weiner and Donahue, "The Doctors Simon Varey and Francisco Hernández."
2. On the impact of the Hernández expedition on Renaissance science, see, for example, Freedberg, *The Eye of the Lynx*, 1–10, 275–304, 456–61.
3. Weiner, "The World of Dr. Francisco Hernández," 3–6; and López Piñero and Pardo Tomás, "The Contribution of Hernández," 124. Naturalists in northern Europe and other regions beyond the reach of the ancients were already beginning to question how comprehensive the ancients' lists were. See Ogilvie, *The Science of Describing*, 60.
4. Myers, *Fernández de Oviedo's Chronicle*, 10, 17, 25, 95–96; and Álvarez Peláez, *La conquista de la naturaleza*, 25–52.
5. Recent scholarship has begun to challenge these tendencies, emphasizing instead a dynamic understanding of cross-cultural exchange. For an insightful discussion of how limitations regarding existing documentation about particular medicines present difficulties for this emerging approach, see De Vos, "Methodological Challenges." Works that attempt to break down the Western/non-Western dichotomy or move beyond teleology include Armus and Gómez, eds., *The Gray Zones of Medicine*; Rodríguez, *Enlightened Immunity*; and De Vos, *Compound Remedies*.
6. On the difficulty of translating concepts, see Burkhart, *The Slippery Earth*; Cervantes, *The Devil in the New World*; and Cooley, "The Giant Remains."
7. Schiebinger, *Plants and Empire*, 87.
8. Rousseau, *Letters on the Elements of Botany*, 3. Note: when quoting from this and other English-language sources, I have modernized the spelling except in cases where the archaic spelling is essential for the analysis.
9. Chabrán, "The Classical Tradition in Renaissance Spain," 10; and Gerbi, *Nature in the New World*, 8–9.
10. There is a rich literature on the functioning and structure of these networks. See, for example, the excellent study by Ogilvie, *The Science of Describing*.
11. Flannery, *In the Herbarium*, 19–23.
12. Rousseau, *Letters on the Elements of Botany*, 4.
13. Kusukawa, *Picturing the Book of Nature*.
14 Dürer, quoted by Pavord, *The Naming of Names*, 17. Note: the literal translation of "*Herbarum vivae eicones*" is "Living Pictures of Plants," but, as Latin specialists Chris van den Berg and Pierre Katsaros have pointed out to me, the figurative understanding would be "Pictures of Living Plants."

15. Vicente Danti, quoted in Kusukawa, *Picturing the Book of Nature*, 8.
16. Dietrich, *Neuw Kreuterbuch*.
17. Oviedo, quoted in Myers, *Fernández de Oviedo's Chronicle*, 69.
18. Ogilvie, *The Science of Describing*, 193.
19. Jiménez de Cisneros, *Complutensian Polyglot Bible*, 25; Castañeda, "Introduction," 8 fn 15.
20. Instructions of Philip II to Dr. Francisco Hernández, January 11, 1570, in Varey, ed., *The Mexican Treasury*, 46; Marroquín Arredondo, "The Method of Francisco Hernández," 45–60.
21. Instructions of Philip II to Dr. Francisco Hernández, January 11, 1570, in Varey, ed., *The Mexican Treasury*, 46.
22. Instructions of Philip II to Dr. Francisco Hernández, January 11, 1570, in Varey, 46.
23. Hernández, "An Epistle to Arias Montano," 629.
24. Álvarez Peláez, *La conquista de la naturaleza*, 25–27, 79; Gerbi, *Nature in the New World*; and Schwartz, *Sea of Storms*, chapter 1.
25. Cortés, "Second Letter," 103.
26. Cervantes de Salazar, *Crónica de la Nueva España*, book 1, 359.
27. Francisco Hernández to King Philip II, Letter 3, November/December, 1571, in Varey, ed., *The Mexican Treasury*, 48–49.
28. Francisco Hernández to King Philip II, Letter 3, November/December, 1571, in Varey, 49.
29. Tezozómoc, *Crónica Mexicana*, 370–71. On whether to attribute this passage to Tezozómoc or to his contemporary indigenous chronicler Chimalpahin, see Townsend, *Fifth Sun*, 228.
30. Cervantes de Salazar, *Crónica de la Nueva España*, book 1, 346.
31. Hernan Cortés, quoted in Granziera, "Huaxtepec," 87.
32. De Vos, "Methodological Challenges," 230–31.
33. Hernández, "An Epistle to Arias Montano," 631.
34. King Philip II, "Instructions to Dr. Francisco Hernández," January 11, 1570, in Varey, ed., *The Mexican Treasury*, 46–47; and Bustamante, "The Natural History of New Spain." Hernández drew on them for his separate study of the antiquities of New Spain; see Varey, ed., *The Mexican Treasury*, 65.
35. Somolinos d'Ardois, *La primera expedición*, 30–37; Lozoya, ed. *Plantas y luces*, 11; Francisco Hernández to King Philip II, Letter 9, March 20, 1575 and "Will of Francisco Hernández," both in Varey, ed., *The Mexican Treasury*, 56, 62; and Hernández, "An Epistle to Arias Montano," 620–34.
36. Hernández, "An Epistle to Arias Montano," 631.
37. Francisco Hernández to King Philip II, Letter 7, March 31, 1573, in Varey, ed., *The Mexican Treasury*, 53.
38. Francisco Hernández to King Philip II, Letter 12, March 24, 1576, in Varey, 59.
39. López-Alonso, *Measuring Up*, 148; Risse, "Shelter and Care," 67.
40. Risse, "Shelter and Care," 65, 72, 76.

41. Francisco Hernández to King Philip II, Letter 12, March 24, 1576; and Francisco Hernández, "Petition to King Philip II," early 1577, both in Varey, ed., *The Mexican Treasury*, 59–60.
42. For a biography of Hinojosos, see Martínez-Hernández, "La llegada del cirujano Alonso López de Hinojosos."
43. Harper, *Plagues upon the Earth*, 119–58.
44. Boruchoff, "Anthropology, Reason, and the Dictates of Faith," 96.
45. Alarcón, *Treatise on the Heathen Superstitions and Customs*, 157–60; and de la Serna, *Manual de ministros de indios*, quoted in Boruchoff, "Anthropology, Reason, and the Dictates of Faith," 96. Francisco Hernández to King Philip II, Letter 4, April 30, 1572; and Letter 6, December 12, 1572, both in Varey, ed., *The Mexican Treasury*, 49–52.
46. Francisco Hernández to King Philip II, Letter 2, May 15, 1571; Letter 3, November/December, 1571; Letter 4, April 30, 1572; and Letter 5, September 22, 1572, in Varey, 48–50.

Chapter 2

1. Francisco Hernández to King Philip II, Letter 9, March 20, 1575, in Varey, ed., *The Mexican Treasury*, 56.
2. On the effects of the 1576 *cocoliztli* outbreak, as well as the earlier 1545 outbreak, see Acuña-Soto et al., "Megadrought and Megadeath."
3. McManus, "Decolonizing Renaissance Humanism," 1161.
4. López Piñero and Pardo Tomás, "The Contribution of Hernández," 123.
5. Álvarez Peláez, *La conquista de la naturaleza*, 111; and Somolinos d'Ardois, *La primera expedición*, 61.
6. López Piñero and Pardo Tomás, *Nuevos materiales y noticias*, 67; Chabrán and Varey, "The Hernández Texts," 6; and Chabrán and Varey, "Entr'acte," 107.
7. López's print shop was located inside the Convento de la Santa Cruz de Tlatelolco, and she sold the book in the shop of Diego Garrido on Calle Tacuba, at the corner with Porteria de Santo Domingo. Safford, "*Lignum Niephriticum*," 227; Emmart, "Introduction," xv; and Chabrán and Varey, "Medical Natural History in the Renaissance," 125. Luigi Guerrini has argued that the John Carter Brown Library's version of Recchi's manuscript likely is the one that had made its way to Mexico where Juan de Barrios (1562–1645) copied it in 1609. Ximénez translated Barrios's text into Spanish, added his own commentary, and published it in Mexico City. Guerrini, *The Accademia dei Lincei*, 6, 18–19.
8. Alcocer, "El mesquite," 691.
9. Guerrini, *The Accademia dei Lincei*, 3.
10. Francisco Hernández to King Philip II, Letter 11, February 10, 1576, in Varey, ed., *The Mexican Treasury*, 57; and Francisco Hernández, "An Epistle to Arias Montano," 632.
11. Francisco Hernández, "The Antiquities of New Spain," in Varey, ed., *The Mexican Treasury*, 77–78.

12. Francisco Hernández, "The Antiquities of New Spain," in Varey, 78. Also see Bustamante, "Natural History," 36; and Lozoya, ed., *Plantas y luces*, 11.
13. De Vos, "Methodological Challenges," 231–32.
14. Varey, "Francisco Hernández, Renaissance Man," 36.
15. López Piñero and Pardo Tomás, "The Contribution of Hernández," 123.
16. Ortiz de Montellano, "El conocimiento de la naturaleza"; and Álvarez Peláez, *La conquista de la naturaleza*, 408–9.
17. Ortiz de Montellano, "El conocimiento de la naturaleza," 117.
18. Francisco Hernández to King Philip II, Letter 3, November/December, 1571, in Varey, ed., *The Mexican Treasury*, 48.
19. Francisco Hernández to King Philip II, Letter 4, April 30, 1572, in Varey, 50.
20. "The Will of Francisco Hernández," in Varey, 62.
21. Francisco Hernández to King Philip II, Letter 2, May 15, 1571, in Varey, 47–48.
22. Boone, "Maps of Territory, History and Community," 112.
23. The original is now housed in the Instituto Nacional de Antropología e Historia, Mexico City, which provides a high-resolution digital facsimile at https://codices.inah.gob.mx/pc/.
24. Gimmel, "Hacia una reconsideración del Códice de la Cruz Badiano"; Robertson, *Mexican Manuscript Painting*, 157; and Laird, "Nahua Latinists."
25. Emmart, "Introduction," 15, 17, 26.
26. Boone, *Descendants of Aztec Pictography*, 47–48.
27. De Gortari, *La ciencia en la historia de México*, 171.
28. Robertson, *Mexican Manuscript Painting*, 44, 47; McDonough, *Indigenous Science and Technology*, 27–77.
29. Emmart, "Introduction," 8.
30. This lack of connection between the text and the images has also been observed by Cooley, "Bezoar," 60.
31. For an insightful and well-placed critique of these tendencies, see Gimmel, "Hacía una reconsideración del Códice de la Cruz Badiano," 276–77.
32. Robertson, *Mexican Manuscript Painting*, 45–46.
33. Dean and Leibsohn, "Hybridity and Its Discontents," 14–15. Regarding the critique of the search for purity, also see McDonough, *Indigenous Science and Technology*, 14, 210.
34. Hassig, "Transplanted Medicine," 32; and Afanador Llach, "Nombrar y representar," 31.
35. Nickelson, "Draughtsmen, Botanists and Nature"; Tobin, "Imperial Designs."
36. Regarding the importance of textual explanations for Renaissance-era botanical illustrations, see Kusukawa, *Picturing the Book of Nature*, 24–25.
37. Burkhart, *The Slippery Earth*; Cervantes, *The Devil in the New World*.
38. Morin González, "Catecismo testeriano"; Robertson, *Mexican Manuscript Painting*, 53–54 and plate 8.
39. Motolinía, quoted in Robertson, *Mexican Manuscript Painting*, 54.
40. Boone, *Descendants of Aztec Pictography*, 25.

41. Williams, "Aztec Soil Knowledge." For other attempts to decipher what these glyphs convey about place, soil, and plant characteristics, also see Emmart, "Introduction," 39–41; Robertson, *Mexican Manuscript Painting*, 157.
42. I intentionally avoid terms such as "hybrid" or "syncretic" to describe the cultural intersections because those terms presume that cultures normally are homogenous, bounded, and in opposition to one another. "Improvisation," by contrast, recognizes human agency and individuals' creative problem-solving. For a useful discussion of hybridity, see Dean and Leibsohn, "Hybridity and Its Discontents."
43. Nieremberg, *Historia naturae maxime peregrinae*; Pimentel, "Baroque Natures," 105; and Varey, ed., *The Mexican Treasury*, xvii–xix.
44. Nieremberg, "Selections from Juan Eusebio Nieremberg's transcriptions of Hernández," in Varey, 174.
45. While there was no equivalent for *yolqui* for plants, later Náhuatl speakers sometimes used the term *quáutl* as a general word for trees. In later centuries, some dialects of Náhuatl broadened this term to describe all plants, but this seems not to have been the case in classical Náhuatl. Ortiz de Montellano, "El conocimiento de la naturaleza," 188.
46. Siméon, *Diccionario de la lengua Náhuatl o Mexicano*.
47. Acosta, *The Natural and Moral History of the Indies*, 183, 184.
48. McManus, "Decolonizing Renaissance Humanism," 1161.
49. Bleichmar and Mancall, "Introduction," 1.
50. Hernández, "An Epistle to Arias Montano," 631.
51. Francisco Hernández to King Philip II, Letter 3, November/December 1571, in Varey, ed., *The Mexican Treasury*, 48.
52. Chabrán, Chamerlin, and Varey, editorial commentary on Francisco Hernández to King Philip II, Letter 3, November/December 1571, in Varey, 48 fn 11.
53. Hernández, "An Epistle to Arias Montano," 631.
54. Francisco Hernández to King Philip II, Letter 12, March 24, 1576, in Varey, ed., *The Mexican Treasury*, 58.
55. Francisco Hernández, petition to King Philip II, early 1577, in Varey, 60.
56. Francisco Hernández to King Philip II, Letter 6, December 12, 1572, in Varey, 51.
57. Council of the Indies to King Philip II, "Sobre las cajas de libros que envía el doctor Francisco Hernández," August 3 and 29, 1576, no. 738, sec. 2, Indiferente General, Archivo General de Indias (hereafter AGI); Viceroy Martín Enríquez to King Philip II, March 30, 1577, Mexico City, fol. 4v., 4th dossier, no. 69, México, AGI; and Council of the Indies to King Philip II, "Sobre la pretensión de el Doctor Francisco Hernández," March 20, 1578, no. 739, Indiferente General, AGI; Baudot, *Utopia and History in Mexico*, 506–7 and fn 39; and Benito-Vessels, "Hernández in Mexico," 48.
58. Baudot, *Utopia and History in Mexico*, 520.
59. Baudot, 508; c.f. López Piñero and Pardo Tomás, who argue that it was simply a matter of Hernández being too sick or old to continue with his work. See López Piñero and Pardo Tomás, "The Contribution of Hernández," 123.

60. Weiner, "The World of Dr. Francisco Hernández," 5.
61. Carrasco and Sessions, *Daily Life of the Aztecs*, 171.
62. Francisco Hernández, petition to King Philip II, early 1577, in Varey, ed., *The Mexican Treasury*, 60.

Chapter 3

1. Cervantes, "Respuesta del discípulo a la carta satisfactoria," 46.
2. Aceves Pastrana, "Las políticas botánicas metropólitanas"; Aceves Pastrana, "La difusión de la química de Lavoisier"; Arias Divito, *Las expediciones*; Fernández Caleya, "Colecciones históricas"; Díaz, "El Jardín Botánico"; González Bueno, "Conocimiento científico"; Lozoya, ed., *Plantas y luces*; Maldonado Polo, *De California a El Petén*; Maldonado Polo, "El primer gabinete de historia natural de México"; McVaugh, *Botanical Results*; Sutherland, "Material tipo de La Colección de Sessé y Mociño"; Rickett, "Royal Botanical Expedition"; San Pío Aladrén and Puig-Samper, eds., *El águila y el nopal*; Sánchez de Lorenzo Cáceres, "Las plantas de la flora mexicana"; Somolinos d'Ardois, *La primera expedición*; Villar, ed., *Homenaje a Martín de Sessé y Juan del Castillo*; and Zamudio Varela, "El peregrinar de las colecciones."
3. Regarding Mexican protonationalism, see Brading, *The First America*; Cañizares-Esguerra, "Postcolonialism *avante la lettre?*"; and Anderson, *Imagined Communities*. Note: a Creole is a person of European extraction born in the Americas. For an insightful critique of the claim that Creole scientific intellectuals of the Enlightenment era were protonationalists, see Labastida, "Ilustración," 31–36.
4. Hall, "Gramsci's Relevance for the Study of Race and Ethnicity."
5. Quoted in Schiebinger, *Plants and Empire*, 7–8.
6. Bustamante, "Natural History," 26; and Lozoya, ed., *Plantas y luces*, 9.
7. Lozoya, ed., 21–35; and Godínez Ortega, "Martín de Sessé," 82.
8. On Banks, see Tobin, *Picturing Imperial Power*, 186.
9. November 23, 1787, folio 323, vol. 138, Reales Cédulas, Archivo General de la Nación, Mexico City (hereafter AGN). My thanks to Rachel Meketon for help with paleography.
10. King Charles III, March 20, 1787, reprinted in Lozoya, ed., *Plantas y luces*, 21.
11. King Charles III, October 27, 1787, reprinted in Lozoya, 30.
12. Sessé y Lacasta, "Expediente sobre los efectos de las plantas medicinales en los enfermos del Hospital de San Andrés," Mexico, 1800, WMS.Amer.44, Microfilm #F2634, Archives and Manuscripts, Wellcome Library Collection, London (hereafter WL).
13. Rodríguez Nozal, "La Oficina Botánica"; González Bueno, "Conocimiento científico"; and González Bueno and Rodríguez Nozal, *Plantas americanas para la España ilustrada*.
14. García Ramos, "Elogio histórico." Several scholars have traced the early details of the personal and professional lives of Sessé and Mociño, as well as other members of the expedition. See, for example, Maldonado Polo, "La Expedición Botánica

a Nueva España," 13–15; Rickett, "Royal Botanical Expedition," 2–6; Olavide, Martínez Shaw, and Puig-Samper Mulero, "José Mariano Mociño," 445–51; and the various essays in Labastida, coord., *La Real Expedición Botánica.* Note: there is disagreement about whether Cervantes was born in 1755 or in 1758.

15. Alvarez ___ and Posada ___ to the Fiscal, July 29, 1787, pp. 16–17, vol. 31, Latin American Collection, William L. Clements Library (hereafter WCL).
16. Note that the number of members varied owing to additions, deaths, and departures. Other scholars have painstakingly reconstructed the expedition's various routes.
17. King Philip II, "Instructions to Dr. Francisco Hernández," January 11, 1570, in Varey, ed., *The Mexican Treasury*, 46–47.
18. Sessé, "Experiencias clinicals y terapéuticas de la Real Expedición Botánica a la Nueva-España," Mexico, 1802, WMS.Amer.43, Microfilm #F2633, WL.
19. For documents related to the debates between Alzate and the members of the Royal Botanical Expedition, see Moreno, ed. *Linneo en México.* Also see Cruz Soto, "Las publicaciones periódicas" and Cañizares-Esguerra, "Postcolonialism *avante la lettre?*," 93–97.
20. Tanck de Estrada, "Aspectos de la intervención."
21. Sessé, "Experiencias clínicas y terapéuticas de la Real Expedición Botánica a la Nueva-España," Mexico, 1802, WMS.Amer.43, Microfilm #F2633, WL. Also see Sessé to Gómez Ortega, draft of a letter, July 3, 1785, V, 1, 1, 3; and "Relación de plantas enviadas al Real Jardín Botánico de Madrid," 1788–1791, V, 2, 6, 2, Archivo Histórico del Real Jardín Botánico (hereafter RJB).
22. Sessé, "Experiencias clínicas."
23. Reeds, "Renaissance Humanism and Botany," 539.
24. Document dated November 23, 1787, folio 323, vol. 138, Reales Cédulas, AGN; Letter, April 26, 1788, 4.[n] 20, RJB; and Arias Divito, *Las expediciones*, 67, 72.
25. Ramírez, "Introduction," v; Rickett, "Royal Botanical Expedition," 5; and García Ramos, "Elogio histórico," 758–59.
26. Ramírez, "Introduction," v; and Aceves Pastrana, "La difusión de la química de Lavoisier." Regarding tensions between the Creole *protomedicato* of the university and the members of the Royal Botanical Expedition, also see Hernández Sáenz, *Carving a Niche*, 34–35.
27. Moncada Maya, *El ingeniero Miguel Constanzó*, 226; and Tanck de Estrada, "Aspectos de la intervención," 191–92.
28. Viceroy Bernardo de Gálvez to the Fiscal, August 31, 1785, pp. 5–8, vol. 31, WCL.
29. Sessé to Viceroy Gálvez, October 14, 1787, pp. 56–58, vol. 31, WCL.
30. Tobin, "Imperial Designs," 276.
31. Royal Order, issued November 11, 1787, recorded March 1, 1788, pp. 76v–92, vol. 31, WCL.
32. Alva ____, Fiscal, to Viceroy, November 3, 1787; Pasava to Fiscal, October 29, 1787, pp. 60–61; Francisco Fernández de Hugo on behalf of the Colegio de San

Gregorio to Viceroy or King, September 26, 1787, pp. 33–50; Cosme de Mier y Trespalacios et al. to Viceroy(?), October 2, 1787, pp. 52–53; and Posava(?), report, October 5, 1787, pp. 54–55, vol. 31, WCL; and Díaz, "El Jardín Botánico," 51. Note: the *desagüe* was a massive drainage engineering project that lasted through most of the colonial era.

33. Sessé to Viceroy, October 14, 1787, pp. 56–58, vol. 31, WCL.
34. Sessé expressed the same view of soil regarding other sites. See Sessé to Viceroy, October 14, 1787, pp. 56–58, vol. 31, WCL. Regarding the transfer of the plot, see Various, pp. 61–71, vol. 31, WCL.
35. Pasava to [Viceroy], March 29, 1788, pp. 121–22, vol. 31, WCL.
36. On the resolution, see Manuel Antonio de Florez to ?, April 2, 1788, p. 122, vol. 31, WCL.
37. Moncada Maya, *El ingeniero Miguel Constanzó*, 200–201. The quotation comes from Fireman and Servin, "Miguel Costanso," 7. Miguel Costanzó later became an instructor of architecture at the Academia de San Carlos and collaborated with Ignacio Castera to oversee construction of the Ciudadela fortress on part of the site that had been designated for the garden.
38. Miguel Constansó, cost estimate, June 19, 1788, pp. 136 and 136v; and annotated site plan, "Mexico City-Botanical Garden," p. 1, vol. 31, WCL; "Plano del edificio del Jardín Botánico que se proyectaba hacer en México," "Plano del Terreno destinado para Jardín Botánico," "Fachada principal que mira á oriente y corte que mira al mediodía á lo largo del edificio sobre la línea a.b.," and "Plano, elevación y perfil de una casa para habitación del Catedrático de Botánica, que debe construirse en el terreno destinado para el jardín botánico de esta Capital de Nueva España," MP-MEXICO, 416, 417, 418, and 419, AGI; Ignacio Castera, "Plano iconografico de la ciudad de México," 1794, G4414.M6 1794.A3, Vault, Geography and Map Division, Library of Congress, Washington, D.C.; Rickett, "Royal Botanical Expedition," 9–12; and Arias Divito, *Las expediciones*, 93–96.
39. Mukerji, *Territorial Ambitions*; Drayton, *Nature's Government*; Sarmiento and González Bueno, "Política científica y expediciones botánicas"; and Thomas, *Man and the Natural World*.
40. Rickett, "Royal Botanical Expedition," 9–10; Díaz, "El Jardín Botánico," 52; and Real Orden de 27 de septiembre de 1791 sobre donativo gracioso, folder 9, vol. 464, Historia, AGN. Regarding cinchona, see Maehle, *Drugs on Trial*.
41. For discussion of the Trasmonte map, see Consentino and Zavala, *Resurrecting Tenochtitlan*, 62. On the desagüe project and the prominence it had during that era, see Candiani, *Dreaming of Dry Land*.
42. For a zoomable version of the Transmonte map, use the stable link: https://www.davidrumsey.com/luna/servlet/detail/RUMSEY~8~1~316594~90085212:Forma-y-Levantado-de-La-Ciudad-de-M#.
43. Cervantes, "La botánica moderna," 145; Arias Divito, *Las expediciones*, 88–90, 105; Rickett, "Royal Botanical Expedition," 9.

44. The loss of these plants and seeds was a major blow and would take five years and much effort and cost to replace. On the effort that went into acquiring the seeds, see Sessé to Gómez Ortega, April 26, 1786, and October 28, 1787, and Sessé to Pedro Acuña y Malvar, January 9, 1794, V, 1, 1, 7, and 12, RJB. Regarding the housing for garden laborers, see "Gastos hechos en el Jardín Botánico establecido en el potrero de Atlampa (el Sapo)," 1791, folder 4, vol. 464, Historia, AGN.
45. Sessé to Intendente General, January (day illegible), 1788, p. 126, vol. 31, WCL. Jacinto arrived via Havana, accompanied by his wife Gertrudis Ramírez. See various letters in folder 3, vol. 464, Historia, AGN.
46. Alcántara Onofre, "The Chinampas of the Valley of Mexico," 9–20.
47. Rickett, "Royal Botanical Expedition," 13–14.
48. See Clavijero, *The History of Mexico*; Alcántara Onofre, "The Chinampas of the Valley of Mexico," 13; Nuttall, *The Gardens of Ancient Mexico*; and Rodríquez Figueroa, *Los jardines nahuas prehispanicos*.
49. Folders 69–70, vol. 464, Historia, AGN; Rickett, "Royal Botanical Expedition," 16.
50. Sessé to Pedro Acuña y Malvar, ministro de Gracia y Justicia, January 9, 1794, V, 1, 4, 30, RJB.
51. Sessé and Cervantes to ?, August 8, 1788, pp. 139v–141v, vol. 31, WCL; Vicente Cervantes to Viceroy Revillagigedo, October 1, 1791, pp. 1–8, folder 8, vol. 464, Historia, AGN; García Ramos, "Elogio histórico," 761; Rickett, "Royal Botanical Expedition," 16.
52. Vicente Cervantes to Viceroy Revillagigedo, October 1, 1791, pp. 1–8, folder 8, vol. 464, Historia, AGN.
53. Vicente Cervantes to Viceroy Revillagigedo, October 1, 1791, pp. 1–8, folder 8, vol. 464, Historia, AGN; Bonilla to Viceroy Revillagigedo, January 28 and February 25, 1791, pp. 1–2, folder 3, vol. 464, Historia, AGN; [illegible] to Viceroy Revillagigedo, December. 31, 1790, p. 3, folder 3, vol. 464, Historia, AGN; Rickett, "Royal Botanical Expedition," 16, 58; [illegible] to Viceroy Félix Berenguer de Marquina, April 6, 1801, folder 3, vol. 464, Historia, AGN; Sessé to Viceroy Félix Berenguer de Marquina, folder 3, vol. 464, Historia, AGN; [illegible] to Viceroy Félix Berenguer de Marquina, January 16, 1803, folder 3, vol. 464, Historia, AGN; and Mascarós to Viceroy Revillagigedo, October 2, 1791, folder 8, vol. 464, Historia, AGN.
54. Vicente Cervantes to Viceroy Revillagigedo, October 1, 1791, pp. 1–8, folder 8, vol. 464, Historia, AGN.
55. Bonilla to Viceroy Revillagigedo, January 28 and February 25, 1791, pp. 1–2, folder 3, vol. 464, Historia, AGN; [illegible] to Viceroy Revillagigedo, 31 December 1790, p. 3, folder 3, vol. 464, Historia, AGN; Rickett, "Royal Botanical Expedition," 16, 58; [illegible] to Viceroy Félix Berenguer de Marquina, April 6, 1801, folder 3, vol. 464, Historia, AGN; Sessé to Viceroy Félix Berenguer de Marquina, folder 3, vol. 464, Historia, AGN; [illegible] to Viceroy Félix Beren-

guer de Marquina, January 16, 1803, folder 3, vol. 464, Historia, AGN. Also see various reports in folder 14, vol. 464, Historia, AGN.

56. Sessé, quoted in Rickett, "Royal Botanical Expedition," 19.
57. Rickett, "Royal Botanical Expedition," 18. The order to shorten the wall likely was never carried out since it would have compromised the security of the viceroy and the royal mint that shared the palace.
58. Vicente Cervantes to Viceroy Revillagigedo, October 1, 1791, pp. 1–8, folder 8, vol. 464, Historia, AGN. On the tension between these two uses in European gardens, see Drayton, *Nature's Government*, 26–49.
59. Sessé, quoted in Rickett, "Royal Botanical Expedition," 17–18.
60. Viceroy Revillagigedo to the Crown, September 30, 1792, quoted in Díaz, "El Jardín Botánico," 73–74.

Chapter 4

1. King Charles III, March 20, 1787, quoted in Lozoya, ed., *Plantas y luces*, 21.
2. Note that there usually is more than one designated type specimen for a species. This occurs for a variety of reasons, including the need for researchers to have a nearby type specimen for use in their research. The most important of the various officially designated samples is its so-called "holotype specimen." A holotype specimen is the sample to which the official Latinate name is attached, usually the first sample of that particular plant ever collected or observed by a scientist operating within the system. This original sample serves as the model against which all other type specimens are judged to confirm whether each of those samples truly is of the same category, or taxon.
3. Note that one of these syntype specimens receives the designation of "lectotype" and serves as the specimen to which the Latinate name is attached and supersedes all other syntypes should any differences be discovered between the lectotype and one or more of the syntypes.
4. Withgott, "Is it 'So Long, Linnaeus'?," 646.
5. While the system is man-made, some have insisted that the aspects of plants that the system uses to organize living organisms are those that are most important within nature. During the Enlightenment era, adherents argued that Linnaeus's system revealed the divine order created by God. More recently, some have come to argue that our systems of organization are based on those features that are most important for evolution or genetics. Whether we see the Linnaean system of organization as a convenient tool or as a revelation of underlying truth, the fact remains that it emerged through a historical process, during which it shunted aside alternative systems and became embedded within power relations.
6. The term comes from Rousseau's critique of pre-Enlightenment botanical practices. Rousseau, *Letters on the Elements of Botany*, 4.
7. Pavord, *The Naming of Names*, 230–39.
8. Pavord, 396.
9. Schiebinger, *Plants and Empire*, 7.

10. Tobin, "Imperial Designs," 273.
11. For an insightful discussion of the importance of visual observation in Linnaean botany, see Bleichmar, "Training the Naturalist's Eye."
12. Tobin, "Imperial Designs," 268. Also see Tobin, *Picturing Imperial Power*, 178–79.
13. Kärin Nickelson uses an illustration of coltsfoot as a compelling example of this practice. See Nickelson, "Draughtsmen," 3.
14. Neri, *The Insect and the Image*, xii.
15. Tobin, "Imperial Designs," 279.
16. Quoted in Nickelson, "Draughtsmen," 9.
17. Jan Kops, quoted in Nickelson, 10.
18. Discussed by Rousseau, *Letters on the Elements of Botany*, 30.
19. Sowerby, *A Botanical Drawing-Book*; and Nickelson, "Draughtsmen," 11. Regarding Linnaeus's emphasis on visuality despite his own arguably limited artistic abilities and for information regarding his collaboration with Ehret, see Charmentier, "Carl Linnaeus."
20. Nickelson, "Draughtsmen," 6.
21. Flannery, *In the Herbarium*, 94.
22. Cervantes, "La botánica moderna," 138.
23. "Instrucción a que deberán arreglarse los Profesores Botánicos, y Naturalistas destinados por S.M. para reconocer las Plantas, Animales y Minerales del Reyno de Nueva España con el general fin de completar e ilustrar los M. SS. y Dibuxos del Doctor Francisco Hernández Proto-médico del Rey Felipe 2°," November 22, 1787, pp. 102–110, vol. 31, WCL.
24. Plan de Enseñanza en la Cátedra que se establece en el Rl. Jardín Botánico de Mexico, November 22, 1787, pp. 93–101v, vol. 31, WCL.
25. "Instrucción que deveran observar los dos Delineadores, o Dibuxantes que de orden de S.M. han de servir con el exercicio de su profesión en la expedición de Botánica, o Historia Natural de Nueva España," November 22, 1787, pp. 110–113v, vol. 31, WCL. Note that the Spanish artists from the Academia de San Fernando in Madrid who accompanied the Ruiz and Pavón expedition to Peru and Chile later received similar instructions, after the Crown created the Oficina Botánica. See Puig-Samper, "Illustrators of the New World," 8–9.
26 Crown to Viceroy and forwarded to the head of the Real Academia de San Carlos, October 27, 1786, pp. 9–10, vol. 31, WCL.
27. Alva ____ to Viceroy, March 23, 1788, pp. 118–20, vol. 31, WCL.
28. Humboldt, *Political Essay*, 214; Widdifield, *The Embodiment of the National*, 15; and Charlot, *Mexican Art and the Academy of San Carlos*, 19–30, 53.
29. Geronimo Antonio Gil and Martín de Sessé to Viceroy(?), April 15, 1788, pp. 128–128v; Joseph Grangido to ?, February 16, 1787, "Mexico City-Botanical Garden," p. 11; Alvarez ____ and Posada ____ to the Fiscal, March 13, 1787, pp. 13–13v; and Alva[rez] ___ to Viceroy, March 23, 1788, pp. 118–20, vol. 31, WCL. Alvaro ___ to the Fiscal Civil, October 27, 1793, exp. IV, vol. 462, Historia, AGN. November 23, 1787, folio 323, vol. 138, and November 23, 1787, folio

323, vol. 138, Reales Cédulas, AGN. Sessé, "Expediente sobre los efectos de las plantas medicinales en los enfermos del Hospital de San Andrés," Mexico, 1800, WMS.Amer.44, Microfilm #F.2634, WL. Letters from Sessé to Casimiro Gómez Ortega, January 27, 1787, March 27, 1787, and November 26, 1787, vol. 1, 1, 10 and 15, RJB.

30. Sessé to Gómez Ortega, June 26, 1788, quoted by both Godínez Ortega, "Martín de Sessé," 100, and Zamudio Varela, "Los artistas," 137.
31. Geronimo Antonio Gil and Martín de Sessé to Viceroy(?), April 15, 1788, pp. 128–128v; and Alvarez ____ and Posada ____ to the Fiscal, March 13, 1787, pp. 13–13v, vol. 31, WCL.
32. Posava to Viceroy, May 9, 1788, pp. 130–130v; and ____ Matamoros (a Regente Fiscal de Rl. Hacienda) to ?, May 19, 1788, pp. 131v–32, vol. 31, WCL; and Vicente de la Cerda, [illegible], and Athanasio Echeverría to conde de Revillagigedo, February 15, 179[final digit illegible], folder 4, volume 462, Historia, AGN.
33. Nickelson, "Draughtsmen," 10.
34. Regarding the paper, tools, and supplies, see Zamudio Varela, "Los artistas," 142.
35. Burdet, "Relato de Augustin Pyramus de Candolle," 99. De Candolle, "Flora de México" is transcribed and included in its entirety on pp. 96–102 of this source. The same transcribed text also appears in de Candolle, "Flora de México," 257–63.
36. Humboldt, *Political Essay*, 216.
37. *Crassulaceae Echeveria gibbiflora* DC, 6331.0463, Torner Collection of Sessé and Mociño Biological Illustrations, Hunt Institute for Botanical Documentation, Carnegie Mellon University, Pittsburgh, PA (hereafter TCHI).
38. *Polemoniaceae*, 6331.1436, TCHI.
39. Sessé, quoted in Zamudio Varela, "Los artistas," 137. Also see Godínez Ortega, "Martín de Sessé," 100.
40. Sessé to Revillagigedo, January 11, 1794, and Sessé to the Fiscal Civil, May 2, 1792, folder 4, vol. 462, Historia, AGN. Also see Olavide, Martínez Shaw, and Puig-Samper Mulero, "José Mariano Mociño," 460–61.
41. Quoted in Bleichmar, "Training the Naturalist's Eye," 19.
42. Regarding ways other Spanish Enlightenment botanical expeditions may have departed from Linnaean expectations, see Bleichmar, "Painting as Exploration."
43. Thanks to Chris Van den Berg for helping decipher the awkward Latin in this note.
44. Nieto, "Presentación gráfica, desplazamiento y aprobación de la naturaleza," 100.
45. It is impossible to reliably distinguish whether particular "N." designations were by the draftsman, the expedition's botanists, or even perhaps sometimes by European botanists after the material was sent to Spain.
46. Original: "*Obserbese la Descripcion al sacarla en limpio para la diseccion de los zamor y flores.*"
47. Sessé and Mociño, "De las plantas," 125.
48. Nickelson, "Draughtsmen," 6.
49. De Candolle, "Flora de México," 99.

50. Quoted in Pavord, *The Naming of Names*, 6–7.
51. Cervantes, "La botánica moderna," 138–40.
52. Sessé and Mociño, "De las plantas," 125.
53. Sessé and Mociño, 125.

Chapter 5

1. Rickett, "Royal Botanical Expedition," 59.
2. Rickett, "Royal Botanical Expedition," 66; "María Guadalupe Morales," Arribadas, 441, N.305, AGI; and Puig-Samper and Rebok, "El reconocimiento." On improvement as a justification for European imperialism, see Drayton, *Nature's Government*. Regarding the number of plant specimens, see Fernández Caleya, "Colecciones históricas," 6.
3. Díaz, "El Jardín Botánico," 76–77; and Fernández Caleya, "Colecciones históricas," 6.
4. Various letters, 1803–1804, Estado 18, no. 96, AGI; and Báez Macías, *Historia de la Escuela Nacional de Bellas Artes*, 71 fn 13.
5. Regarding Mociño's accomplishments in Spain, see Labastida, "Ilustración," 31–49.
6. Puig-Samper and Rebok, "El reconocimiento," 6–7; McVaugh, *Botanical Results*, 12–13; Olavide, Martínez Shaw, and Puig-Samper Mulero, "José Mariano Mociño," 455.
7. Augustin Pyramus de Candolle, quoted in McVaugh, *Botanical Results*, 13. Regarding the loss of the manuscripts, many of which accidentally ended up in Paris, see De Candolle, "Flora de México," 98.
8. De Candolle, 100.
9. De Candolle, 100.
10. Olavide, Martínez Shaw, and Puig-Samper Mulero, "José Mariano Mociño," 456.
11. Fernández Caleya, "Colecciones históricas," 6.
12. Fernández Caleya, 7; and McVaugh, *Botanical Results*, 1–2.
13. Sessé and Mociño, *Plantae Novae Hispaniae* and Sessé and Mociño, *Flora Mexicana*.
14. Fernández Caleya, "Colecciones históricas," 8.
15. November 23, 1787, folio 323, vol. 138, Reales Cédulas, AGN; and Letter, April 26, 1788, in Arias Divito, *Las expediciones*, 72; Rickett, "Royal Botanical Expedition," 66; "María Guadalupe Morales," Arribadas, 441, N.305, AGI; and Puig-Samper and Rebok, "El reconocimiento."
16. Humboldt, *Political Essay*, 216.
17. García Ramos, "Elogio histórico," 761.
18. Cervantes to Viceroy Francisco Javier Venegas de Saavedra, April 2, 1811, folder 8, vol. 462, Historia, AGN; and Rickett, "Royal Botanical Expedition," 57, 67.
19. Cervantes, quoted in Rickett, "Royal Botanical Expedition," 67.
20. Velasco quoted in Rickett, 68. He might be referring to Diego Martín de la Cruz, a long-serving assistant at the garden, or perhaps to a different employee with a similar name.

21. Rickett, 68–69. On the conflict between Velasco and Cervantes, see various letters, folder 8, vol. 466, Historia, AGN.
22. Various letters, folder 18, vol. 466, Historia, AGN.
23. Various, April 1818–July 1818, folder 17, vol. 466, Historia, AGN; Rickett, "Royal Botanical Expedition," 69; Cervantes to unknown, June 14, 1821; Cervantes to unknown, June 16, 1821; Secretary of the Viceroy, 1821; Jose Maldonado to unknown, June 21, 1820; Mariano Lopes to unknown, September 18, 1820; and various, folders 31–32, vol. 466, Historia, AGN.
24. Ramírez, "Reseña de la Expedición de Historia Natural"; García Ramos, "Elogio histórico"; "Expedición al Virreinato de Nueva España."
25. Bullock, *Travels in Mexico*, 183–84.
26. Bullock, 184–85.
27. Bullock, 184–87.
28. Lucas Alamán, "Épocas de los principales sucesos de mi vida," August 28, 1843, series 1, box 1, exp. 236, p. 5, Lucas Alamán Papers, 1598–1853, Benson Latin American Collection; Bullock, *Travels in Mexico*, 183–86. Lizardi, report, box 10, folder 2, 1821; Cervantes to exmo Señor, August 27, 1823, box 49, folder 16/1, 1823; correspondence of Lucas Alamán, various, 1823–1829; congressional decree, November 21, 1831, box 411, folder 1; all from Gobernacion "sin sección," AGN, Mexico City. Thanks to Eric Van Young for introducing me to the papers of Lucas Alamán. On the Hospital de Indios (Naturales), see Romero-Huesca and Ramírez Bolla, "La atención médica," especially p. 497.
29. García Ramos, "Elogio histórico," 765.
30. León, *Botánico-Mexico*, 78; Ibarra Cabrera, "Profesor Don Miguel Bustamente y Septiem." On the garden as a site for social prestige among the new elite, see Vega y Ortega, "El desarrollo del jardín botánico."
31. Quoted in Rickett, "Royal Botanical Expedition," 70.
32. Sociedad Mexicana de Geografía y Estadística, *Biografía del Señor don Benigno Bustamante*, 4; Maldonado, "La vida y la obra de Don Miguel Bustamante"; and Vega y Ortega, "'Una ciencia tan útil como agradable.'"
33. Alcocer, *El bosque de Chapultepec*, 1, available through the Wellcome Collection, London, England. For more on the Sociedad Mexicana de Historia Natural, see chapters 6–8.

Chapter 6

1. Alcocer, *El bosque de Chapultepec*, 317.
2. Lucas Alamán, Secretario de Relaciones Interiores y Exteriores, added the word "nacional" at the last minute as a condition for government funding. Sánchez Hernández, "Instituto de Ciencias, Literatura y Artes."
3. Before 1867, the museum was narrowly focused on gathering and inventorying items, but after the Liberal triumph over Maximilian it became a center for "the conservation of the origins" of the Mexican Republic. See Morales Moreno,

Origenes de la museologia mexicana, 38. Regarding the original spelling, see Rutsch, "Natural History, National Museum and Anthropology in Mexico," 94.

4. This commonality was noted by Torres Montero "Nature and Culture in Nineteenth-Century Mexico," 14.
5. Mexico was preceded only by France (1821), Germany (1828), and England (1830). Spain, interestingly, did not follow suit with its own geographical society until 1876. Because of political disruptions, the chartered institute was not created until January 26, 1835. During its early years, the SMGE underwent various name changes, and from 1839 to 1849 it was part of the Ministerio de Guerra y Marina. See Treviño Urquijo, *Comisión Geográfica-Exploradora*, 22–23.
6. Though documents from the time refer to it by a range of names, I use the term "Secretaría de Fomento" except when quoting historical documents that might use a different term.
7. The SMGE's mapping efforts have been well studied by Craib, *Cartographic Mexico*; Carrera, *Traveling from New Spain to Mexico*; and Moncada Maya and Gómez Rey, "Patrimonio geográfico mexicano."
8. Arriaga, "Informe," 263; and "Prospecto." On funding, see Bárcena, "Informe," 266.
9. Editors, "Introduction," v; García Cubas, "Reseña de los trabajos"; Payno, "Introduction," iii; Azuela Bernal, *Tres sociedades científicas*, 21–25; García Cubas, "Sociedad de Geografía y Estadística," 2; Arroyo, "Que hace la Sociedad de Geografía y Estadística"; "Descripción de la cochinilla mixteca"; Larrainzar, "Noticia histórica de Soconusco"; and Escobar, "Noticias geográficas."
10. See, for example, Hill, "Cera vegetal"; Gonzalez, "Algodón"; Rivera y Mendoza, "Informe." Also see González Cos, "Estadística del partido de Silao"; Hernández y Dávalos, "Estado de Jalisco"; Covarrubias, "Apuntes estadísticos"; Pumian, "Noticia geográfica y stadística"; and García, "Producciones utilísimas."
11. Craib, *Cartographic Mexico*, 36; and Hale, *Liberalism in the Age of Mora*, 217.
12. Pani, "Los viajeros decimonónicos," 32.
13. Payno, "Notas geologicas."
14. Aldherre, "Los indios de Yucatán," 74.
15. De Fleury, "Notas geológicas," 115.
16. Gagern, "Rasgos característicos," 802. Gagern had fled Berlin in late 1852 or early 1853 to the United States, where he was struck by the beauty of the natural landscape but was disappointed by the culture of the people. After traveling from Niagara Falls to New Orleans, he met Manuel Payno, who gave him a letter of introduction that helped him secure a job as a military instructor at the Colegio Militar de Chapultepec. He introduced Prussian military methods, became a captain in the corps of engineers, and soon gained Mexican citizenship. For Gagern's biography, see Celis de la Cruz, "Presencia de la masonería alemana."
17. Gagern, "Rasgos característicos," 802–18.
18. Mendoza, "Discurso"; and Carrillo, "Disertacion sobre la literature antigua." Mendoza's claim resonates with what historian Marisol de la Cadena found

regarding the study of Quechua in Peru during the early twentieth century. See de la Cadena, *Indigenous Mestizos*, 76.

19. Ruiz, "Riqueza de Michoacán."
20. Regarding this aspect of Linnaean binomials, see Schiebinger, *Plants and Empire*, 194–203.
21. Garcia Cubas, et al., "Dictamen," 602; Romero, "Reseña de los trabajos científicos," 75; and Mendoza, "Discurso." For an insightful analysis of the fixing of place-names as a means of stabilizing historical and cultural associations, see Brown, *Inventing New England*, 57–73.
22. Garcia Cubas, et al., "Dictamen," 602.
23. Payno, "Anáhuac-México-Tenochtitlan-América," 206.
24. Mendoza and Herrera, "El anacahuite," 151.
25. Mendoza and Herrera, 151.
26. See, for example, Riva Palacio, *México a través de los siglos*; Garrigan, *Collecting Mexico*; and Widdifield, *The Embodiment of the National*.
27. Pimentel, *Memoria*; and Pimentel, "Observaciones," 235.
28. Casas, "Botánica, raiz de jalapa," 6. Note: I have been unable to learn who Casas was outside his participation in the SMGE.
29. Ruiz, "Riqueza de Michoacán," 324–25.
30. Escobar, "Noticias geográficas."
31. Mendiondo, "El Té," 525. Note: I have been unable to learn who Mendiondo was outside his participation in the SMGE.
32. See Brailovsky and Foguelman, *Memoria verde*; and Dunlap, *Nature and the English Diaspora*.
33. Bárcena, "Aclimatación de plantas," 141. Also see Bárcena, "Observaciones de plantas características."
34. Martins, "Las poblaciónes vegetales."
35. Payno, "El Maguey Mexicano."
36. Fernández, "Analysis del mezcal."
37. Alejo Torres, "Instrucción sobre el cultivo del cacao," 106.
38. Reyes, "Ligeras nociones," 119.
39. Crosby, *The Columbian Exchange*.
40. Herrera, "Revista científica: El plátano."
41. Ramírez, "Lecturas de historia política de México," 73. For a historical source from 1598 detailing the process by which plants were transported to elites' gardens from the Aztec empire, see Tezozómoc, *Crónica Mexicana*, 370–71.
42. Arriaga, "Informe," 272–73.
43. Alcocer, "El bosque de Chapultepec," 317. Also see Alcocer, *El bosque de Chapultepec*.
44. Alcocer, "El bosque de Chapultepec," 318.
45. Alcocer, 318–19.
46. Alcocer, 320–22.
47. The first installment was Herrera, "Sinonimia vulgar y científica."

48. "Catálogo de la colección de productos naturales indígenas"; "Catálogo de la colección de Drogas indígenas"; and Cordero y Hoyos, "Herbario de plantas indígenas."
49. Villada, "Crónica"; González, "La florula de la ciudad de Monterrey"; and Bárcena, "El Jardín Botánico." Shelley Garrigan has shown that lists and statistical tables were as powerful as maps or canons of art and archaeology in affirming the existence of the nation. They also affirmed the ability of scientific, intellectual, and political elites to steadily fill in the blanks through rational and firm management of their territory, as they guided Mexico toward a future of modernity, progress, and order. In this way, the proliferation of lists and statistical tables—including incomplete ones—became a "way of viewing and displaying the unruly and defiant national landscape as a manageable possession." See Garrigan, *Collecting Mexico*, 154.
50. Regarding the 1871 discovery of the manuscripts that had been thought lost, see Peñafiel, *La Naturaleza* addendum at the end of issue 22, vol. 2 (1873), no page number.
51. González, "Discurso sobre el Estudio de la Botánica," 179–80. Note: González also was a founder of the Universidad Autonoma de Nueva León and a hospital in Monterrey that later took his name as the Hospital Universitario Dr. José Eleuterio González.
52. Morales, "El ypacihuitl."
53. Herrera, "Apuntes para la materia médica," 187.
54. Bárcena, "Discurso," 4.
55. Clavijero, "Breve noticia."
56. Clavijero, "Disertacion IV"; Alzate y Ramírez, "Botanica"; and Alzate y Ramírez, "Noticia de algunas plantas." Regarding eighteenth-century Mexican refutations of Buffon and critiques of Linnaeus, see various documents in Moreno, ed., *Linneo en México*; Aceves Pastrana, "La difusión de la química de Lavoisier"; Cruz Soto, "Las publicaciones periódicas"; and Cañizares-Esguerra, "Postcolonialism *avant la lettre*?" See also Alzate y Ramírez, *Gacetas de literatura*, especially vol. 3.
57. For a compelling critique of Mociño as a protonationalist, see Labastida, "Ilustración."
58. Mociño, "Discurso."
59. Note: at the time of Mociño's speech, his calls seem to have had an effect, as indicated by the bylaws of the Mexican Pharmaceutical Academy, which placed near the top of its priority list the creation of a Mexican pharmacopeia. See Academia Farmacéutica de México, *Reglamento de la Academia Farmacéutica*, 3, available through the Wellcome Collection, London, England.
60. Sosa, "Don José Mariano Mociño." Also see Ramírez, "Reseña de la Expedición de Historia Natural," which establishes the official narrative of the Royal Botanical Expedition as part of the lineage culminating in the endeavors of the SMHN.
61. Sosa, "Don Miguel Bustamante," 126–27.

62. The list first appeared in print in 1887 as *Plantas de Nueva España coleccionadas por mandato y bajo los auspicious del benignisimo Rey Carlos IV, y ordenadas conforme al system de Linneo*. That same year, the SMHN also published Sessé and Mociño, *Flora Mexicana* and Sessé and Mociño, *Plantae Novae Hispaniae*.
63. Noriega, "El ahuehuete"; Ortega Reyes, "El gigante de la flora mexicana," 110; "¿Cuál debe ser el árbol nacional?"; and "Se anima el concurso para elegir el árbol nacional."
64. Bárcena, "El árbol de manitas," 114–16; Peñafiel, "Apuntes de viaje."
65. Bárcena, "El árbol de manitas," 116–17.
66. Bárcena, "La hauya elegans," 307.
67. This effort to remember and recover is, to some extent, a contrast to the Enlightenment practice of agnotology, discussed previously.

Chapter 7

1. Quotation from Bárcena, "Discurso," 4.
2. De la Llave, "Memoria acerca."
3. Mociño, "Descripción del volcán de Tuxtla."
4. Burke, "A Philosophical Inquiry," 277–83; Cronon, "The Trouble with Wilderness," 70–71; and Stott, "Jungles of the Mind," 38–44.
5. Thoreau, "Walking," 657–74; and Oelschlaeger, *The Idea of Wilderness*, 165.
6. Malanco, "Discurso pronunciado," 149.
7. Rovirosa, "Recuerdos."
8. Rovirosa, 271.
9. See, for example, Cronon, "The Trouble with Wilderness," 69–90.
10. Rovirosa, "Recuerdos," 272–73.
11. Regarding Humboldt's approach to nature, see Wulf, *The Invention of Nature*.
12. Pratt, *Imperial Eyes*, 39, 61.
13. Villada, "Breve noticia."
14. Villada, 558.
15. Potts, "Chihuahua." Regarding Potts's role in Chihuahua, see Webb, *Adventures in the Santa Fé Trade*.
16. Potts, "Chihuahua," 282.
17. Regarding the importation of exotic species for hunting in Australia and Argentina, see Brailovsky and Foguelman, *Memoria verde*, 206–11; and Dunlap, *Nature and the English Diaspora*. On hunting as a pathway into nature appreciation in the U.S., see Leopold, *Sand Country Almanac*.
18. Potts, "Chihuahua," 282.
19. Regarding Ramírez, see Sarabia, *La consolidación de la botánica mexicana*, 63. Morales quotes at length Ramírez's account of his ascent of Popocatepetl on pp. 80–81.
20. Altamirano Piolle, *José María Velasco*, 13; and Oles, *Art and Architecture in Mexico*, 43. Note: because Altamirano had access to Velasco's archives, her biography of the artist stands out as the most thorough and insightful to date.

21. Altamirano Piolle, *José María Velasco*, 29–41, 58–59, 86, 100; Phaf-Rheinberger, "Between History of Art and History of Science," 248.
22. Though *casta* paintings of the eighteenth century contained imagery that superficially might appear related to natural history, neither Velasco nor any of Mexico's leadings naturalists of the nineteenth century—indeed, not even earlier naturalists, such as those of the Royal Botanical Expedition—took any interest in these paintings. It is possible that naturalists of the eighteenth and nineteenth centuries found casta paintings' dubious categorizations lacking in scientific rigor. As historian Douglas Cope has shown, the numerous racial categories portrayed in casta paintings bore little relationship to the reality of how race was understood or experienced in people's lives within Mexico. See Cope, *The Limits of Racial Domination*. Another possible explanation for their lack of mention by Velasco and other naturalists is that casta paintings, which were created for foreign audiences, might have been poorly known within Mexico before the cultural renaissance that followed the Mexican Revolution, when collectors took a lively interest in previously neglected art forms.
23. Altamirano Piolle, *José María Velasco*, 100.
24. Landesio, *La pintura general o de paisaje y la perspectiva en la Academica Nacional de San Carlos* (Mexico, 1867), 6–7, quoted in Altamirano Piolle, *José María Velasco*, 91.
25. On the Hudson River School, see Driscoll, *All That is Glorious*.
26. Altamirano Piolle, *José María Velasco*, 130, 148, 219.
27. Velasco, "Discurso," 4–9.
28. Altamirano Piolle, *José María Velasco*, 139.
29. Velasco to his family, July 4, 1889, Paris, in Altamirano Piolle, 310. Altamirano Piolle transcribes in their entirety the letters that Velasco wrote; see pp. 296–317.
30. Velasco, quoted in Altamirano Piolle, 322–23.
31. Sherman and Meyer, *The Course of Mexican History*, 405–7.
32. Haber, *Industry and Underdevelopment*, 15; and Schmidt, "The Social and Economic Effect."
33. On the "tropicalization" of Latin America, see Stepan, *Picturing Tropical Nature*, 18–36. Regarding Church, see, for example, his *Scene on the Magdalena* (1854), *South American Landscape* (1856), and *View of Cotopaxi* (1867).
34. Altamirano Piolle, *José María Velasco*, 399–417.
35. Craib, *Cartographic Mexico*, 37, 39.
36. García Cubas, *Atlas pintoresco é histórico*. Available in high resolution from the Library of Congress at the stable link https://www.loc.gov/item/2008621671/.
37. Villada, "Apuntes de geología y de botánica," 419.
38. Puga, "El Cerro de la Estrella," 488.
39. Herrera, "Informe," 130.
40. Simonian, *Defending the Land of the Jaguar*, 56; and Tenorio-Trillo, *Mexico at the World's Fairs*, 16.
41. Instituto Médico Nacional, *Datos para la materia médica*, 5–6.

42. Turner, "The Significance of the Frontier," 31–60. His frontier thesis also offered a rosy nationalist narrative separate from the burden of Southern history rooted in black slavery.
43. Sessé and Mociño, *Plantae Novae Hispaniae*, 2.
44. Despite some interest in fauna, animals did not attract the same level of interest that flora did, neither during the Enlightenment nor during the nineteenth century. This possibly tied to the fact that most—but not all—medicinal simples derived from plants, rather than from animals. It likely also can be traced to the lingering effects of the Renaissance and Nahua notions about the different relationship plants had to the land, as compared to that enjoyed by peripatetic creatures that traveled by foot, fin, or wing (see chapter 2). Whatever the reasons might have been, plants were thought to be more rooted to place and nation than were animals.
45. Alcocer, "Reseña histórica."
46. Alcocer, 13. One could argue that this effort paved the way for the 1940s creation of the birth control pill using an indigenous wild yam, the history of which is recounted by Soto Laveaga, *Jungle Laboratories*.
47. Alcocer, "Reseña histórica," 13–16.
48. Sosa, "Nuestro Programa," *El estudio semanario de ciencia médica* 1, no. 1 (June 10, 1889): 1–2, quoted in Sarabia and Pastrana, "Datos para la materia médica," 15.
49. Instituto Médico Nacional, *Datos para la materia médica*, vol. 1, 5–6. Also see Sarabia and Aceves Pastrana, "Datos para la materia médica"; Sarabia, *La consolidación de la botánica mexicana*, 43; and Schifter Aceves and Sarabia, "La trayectoria de Francisco Río de la Loza." A total of five volumes appeared at regular intervals through 1908 (the year of Altamirano's untimely death). Though it was the Paris exhibition that prompted the creation of the IMN, the institute's roots traced back to the SMHN's project, launched by Arriaga back in 1873, to develop a botanical geography of the nation. It also was an extension of the project by *La Naturaleza* from 1873 through 1883 to identify and study the potential of native medicinal plants (see chapter 6).
50. Sarabia and Aceves Pastrana, "Datos para la materia médica," 17–18.
51. Instituto Médico Nacional, *Datos para la materia médica*, part 1, 14.
52. Instituto Médico Nacional, part 1, 10.
53. Instituto Médico Nacional, part 1, 12.
54. García Barragán, "El paisajista Adolfo Tenorio." García Barragán suggests that Tenorio's work is little known today because almost all of it is held in private collections, unavailable for public viewing or academic study.
55. Instituto Médico Nacional, *Datos para la materia médica*, part 1, plate between pp. 204 and 205.
56. Sarabia and Aceves Pastrana, "Datos para la materia médica," 24–28.
57. Sarabia and Aceves Pastrana have found that, despite the aspirations of Altamirano, Sosa, and other researchers, the IMN failed to get the publication into the hands of many rural doctors. In 1896, Francisco Río de la Loza set out to sell just

one hundred copies of volume one domestically and was not even able to accomplish this because of the lack of a system for promotion and distribution in rural regions. Later volumes fared even worse. Sarabia and Aceves Pastrana, 22–23.

58. Instituto Médico Nacional, *Datos para la materia médica*, part 1, 28.
59. Regarding Diego Rivera and postrevolutionary nationalist aesthetics, see López, *Crafting Mexico*.
60. Lekan, *Imagining the Nation in Nature*, 8.

Chapter 8

1. Balbontín, "El lago de Texcoco," 373.
2. Ruiz, "Riqueza de Michoacán," 321–25.
3. Zárate, "El Estado de Veracruz en 1897," 33.
4. Zárate, 34–35.
5. Dean, *With Broadax and Firebrand*, 231; Das, *Colonialism, Development, and the Environment*; Brailovsky and Foguelman, *Memoria verde*, 196–206; and Green, *The Spectacle of Nature*.
6. Castro, *Mexico y sus alrededores*.
7. Castro, Signone, and García Cubas, *Álbum del ferrocarril*.
8. Altamirano, "Crónica de las excursiones científicas," 125–26.
9. Balbontín, "Los bosques," 145–46.
10. Editors, "Selvicultura," 9.
11. Editors, 9.
12. Gil, "Destrucción de los bosques," 12.
13. Editors, "Selvicultura," 9.
14. Editors, 12–13.
15. Salonio, "Ordenanzas antiguas sobre bosques," 15.
16. García Cubas, "Reseña de los trabajos," 934. To this end, the SMGE issued frequent proposals for action ("*excitativas*") to government officials and formed its own commission to study and protect Mexico's forests for the good of the nation.
17. Ramírez, et al., "Bosques y arbolados," 17. Note: Gumesmindo Mendoza was a member of both the SMGE and the SMHN. See Torres Montero, "Nature and Culture in Nineteenth-Century Mexico." I have been unable to locate information about Ignacio Cornejo.
18. Ramírez, et al., 21–22. For an outstanding study of what really was unfolding in France during these years, see Whited, *Forests and Peasant Politics*. For an overview of Mexican forest conservation regulations during the nineteenth century, see Simonian, *Defending the Land of the Jaguar*, 49–66. Note that despite his many insights, Simonian mistakenly assumes that the SMGE's prescription was the consensus among Mexico's intellectuals. He also exaggerates Mexicans' utilitarian view of forests and erroneously accepts the SMGE's claim that peasants were the primary culprits.
19. Ramírez, et al., "Bosques y arbolados," 14, 15, 20.
20. Ramírez, et al., 23–24.

21. Davis, *Resurrecting the Granary of Rome*; and Whited, *Forests and Peasant Politics*, 201–3.
22. Schmidt, "The Social and Economic Effect," 31–32; Coatsworth, *Growth Against Development*, 154–55; and Hernández, *Bad Mexicans*, 34–35.
23. Payno, "Bosques y arbolados," 81.
24. Payno, 85, 88.
25. Payno, 90.
26. Payno, 81, 85.
27. Balbontín, "Los bosques," 144. Note: "Nation" is capitalized by Balbontín. Regarding his military career, see Balbontín, *Memorias del Coronel Manuel Balbontín*.
28. In this instance, Balbontín is quoting the Porfirian-imposed governor of Guerrero, Francisco O. Arce, who was notorious for expropriating land from indigenous peasants and redistributing it to the new bourgeoisie to build a base of support for Porfirio Diaz. Balbontín, "Los bosques," 144–45. For more on Arce's role in Guerrero, see López, *Crafting Mexico*, 212–13.
29. Balbontín, "Los bosques," 146–47.
30. Finck, "Una excursión a las faldas," 234.
31. Balbontín, "Los bosques," 147, 151.
32. Herrera, "Informe acerca de los trabajos," 130.
33. Schwarz and Jax, *Ecology Revisted*, 102–103, 145–47. U.S.-based biologists tend to presume that insights into ecology and ecosystems historically came almost exclusively from U.S. scientists, and, to a lesser degree, other Anglophone practitioners. They thus ignore or minimize how the concepts were embraced or deployed beyond the U.S. before World War II. See, for example, Golley, *A History of the Ecosystem Concept in Ecology*.
34. Bárcena, et al., "Dictámen sobre le repoblación vegetal," 145.
35. Villada and Armendariz, "Necesidad de la conservación de los bosques," 3–10.
36. Bárcena, "El bosque de Chapultepec," 193–98.
37. Altamirano, "Necesidad de la repoblación de los bosques," 11.
38. Altamirano, 13. The claim about the introduction of new species was likely meant to quell growing criticism of the use of eucalyptus in reforestation projects.
39. Altamirano, 13.
40. Altamirano, 15, 18, 21.
41. Sánchez, "Fundación del Museo Nacional de Historia Natural," 1–6.
42. Treviño Urquijo, *Comisión Geográfica-Exploradora*, 70, 96, 130–36.
43. Villada, "Obituary of Manuel Urbina y Altamirano." These losses included Manuel Urbina y Altamirano, Antonio del Castillo, Joaquin Arriaga, Mariano Bárcena, Alfonso Herrera, José Rovirosa, José Ramírez, José Segura, Gumesindo Mendoza, Leopoldo Rio de la Loza, Alfredo Dugés, Miguel Pérez, and José María Velasco.
44. Treviño Urquijo, *Comisión Geográfica-Exploradora*, 96, 136.
45. Rutsch, "Natural History, National Museum," 103. The university had originally been founded in 1551 but had ceased operations under Maximilian and was revived and restructured in 1910.

46. Wakild, *Revolutionary Parks*, 13, 45.
47. This botanical revival was similar to what historian Stuart McCook observed regarding the populist 1930s in the Spanish Caribbean. McCook, *States of Nature*.
48. *Enciclopedia nacional popular*, which is analyzed in López, *Crafting Mexico*, 143–44.
49. Simonian, *Defending the Land of the Jaguar*, 67–110; Wakild, *Revolutionary Parks*, 27, 60–69, 139–41; Wakild, "Parables of Chapultepec," 198–200; and Boyer, *Political Landscapes*, 55–57.
50. Radkau, "Germany as a Focus of European 'Particularities,'" 20–21.

Conclusion

1. López, *Crafting Mexico*. This effort extended even into the urban center of the capital. Art historians Delia Consentino and Adriana Zavala have shown, for example, that it was not until the 1920s, in the midst of a massive postrevolutionary urban migration which more than tripled the city's population, that the masses were encouraged to ponder their own direct ancestors' role in building "the bridges, causeways, huts, and temples of Tenochtitlán." Consentino and Zavala, *Resurrecting Tenochtitlan*, 5.

Postscript

1. Florescano, *La bandera Mexicana*, 25–64, 71–74, 128–39, 154. One of the oldest existing Aztec representations of the eagle on the nopal comes from the stone carving known as the Teocalli de la Guerra Sagrada, unearthed by archaeologists near the Palacio Nacional in 1926. For an insightful analysis of that object and of similar depictions in native manuscripts, see Boone, "Migration Histories as Ritual Performance," 139–40.
2. López, *Crafting Mexico*, 185–92.
3. Quoted in DeParle, "Next Generation of Conservative."
4. Bye, "The Role of Humans"; and Edwards, "The Human Impact on the Forest."
5. Bray and Klepeis, "Deforestation, Forest Transitions, and Institutions for Sustainability." Also see the essays in Bray, Merino, and Barry, eds., *Los bosques comunitarios de México*.
6. Beltrán, "Use and Conservation"; UNESCO, "Task Force on Criteria and Guidelines"; Smardon and Faust, "International Policy in the Biosphere Reserves." Thanks to Winslow Dahlberg-Wright for first revealing to me the connections between Beltrán's original proposal and the ultimate UNESCO program.
7. Matchett, "At Odds over Inbreeding"; and Bravo, "Abren banco de semillas."
8. Torres, "Sheinbaum formaliza su reforma"; and "Mexican President Confident."
9. See, for example, Sosa-Escalante and Álvarez, "El estudio de la biodiversidad en México."
10. This has been the case even in high-profile conflicts over the Lacandon forest and the Montes Azules Biosphere Reserve in Chiapas. See, for example, Simon, *Endangered Mexico*, 91–125, and O'Brien, *Sacrificing the Forest*, 139–84.

ARCHIVES OF UNPUBLISHED DOCUMENTS

Archivo General de Indias (AGI), Seville, Spain.
    Estado
    Indiferente General
Archivo General de la Nación (AGN), Mexico City, Mexico.
    Reales Cédulas
    Historia
    Gobernación
Archivo Histórico del Real Jardín Botánico (RJB), Madrid, Spain.
Benson Latin American Collection, University of Texas at Austin, Austin, TX.
    Lucas Alamán Papers, 1598–1853
Biblioteca Nacional de Antropología e Historia, Mexico City, Mexico.
    Collección de Códices
Bibliothèque National de France, Paris, France.
    Départment des Manuscrits, Mexicain
Hunt Institute for Botanical Documentation, Carnegie Mellon University, Pittsburgh, PA.
    Torner Collection of Sessé and Mociño Biological Illustrations (TCHI)
Library of Congress, Washington D.C.
    Geography and Map Division
Wellcome Library (WL), London, England.
    WMS/Amer, Archives and Manuscripts
William L. Clements Library (WCL), University of Michigan, Ann Arbor, MI.
    Latin American Archives Collection

BIBLIOGRAPHY OF PUBLISHED MATERIAL

Aboites, Luis. "Relación sociedad-naturaleza desde la historia de los usos del agua en México, 1900–1940." In *Estudios sobre historia y ambiente en América, Volume 1: Argentina, Bolivia, México, Paraguay*, edited by Bernardo García Martínez and Alba González Jácome. El Colegio de México, 1999.

Academia Farmacéutica de México, *Reglamento de la Academia Farmacéutica de México*. Emprenta de la Testamentaría de Valdéz, 1839.

Aceves Pastrana, Patricia. "La difusión de la química de Lavoisier en el Real Jardín Botánico de México y en el Real Seminario de Minería (1788–1810)." *Quipu* 7, no. 1 (1990): 5–35.

Aceves Pastrana, Patricia. "Las políticas botánicas metropólitanas en los virreynatos de Nueva España y de Perú." In *Mundialización de la ciencia y cultura nacional*, edited by A. Lafuente, A. Elena, and M. L. Ortega. Doce Calles, 1993.

Acosta, Joseph (José) de. *The Natural and Moral History of the Indies*. Translated by Edward Grimston. 1604, reprinted Hakluyt Society, 1880.

Acuña-Soto, Rodolfo, David Stahle, Malcolm Cleaveland, and Matthew Therral. "Megadrought and Megadeath in 16th century Mexico." *Emerging Infectious Diseases* 8, no. 4 (2002): 360–62.

Afanador Llach, María José. "Nombrar y representar: Escritura y naturaleza en el Códice de la Cruz-Badiano, 1552." *Fronteras de la Historia* 16, no. 1 (2011): 13–41.

Alarcón, Hernando Ruiz de. *Treatise on the Heathen Superstitions and Customs That Today Live Among the Indians Native to This New Spain, 1629*. Translated by Ross Hassig. University of Oklahoma Press, 1984.

Alcántara Onofre, Saúl. "The Chinampas of the Valley of Mexico." In *Gardens and Cultural Change: A Pan-American Perspective*, edited by Michel Conan and Jeffrey Quilter. Dumbarton Oaks, 2007.

Alcocer, Gabriel. *El bosque de Chapultepec: Proyecto de un Jardín Botánico*. Mexico, 1886.

Alcocer, Gabriel. "El bosque de Chapultepec, proyecto de un jardín botánico." *La Naturaleza* 7 (1887): 317–23.

Alcocer, Gabriel. "El mesquite." *La Naturaleza* series 2, vol. 3, (1903): 683–95.

Alcocer, Gabriel. "Reseña histórica de los trabajos emprendidos acerca de la flora mexicana." *La Naturaleza* appendix to series 2, vol. 3 (1903): 11–24.

Aldherre, Federico. "Los indios de Yucatán (1865)." *Boletín de la Sociedad Mexicana de Geografía y Estadística* 2a época, vol. 1 (1869): 73–81.

Alejo Torres, Leon. "Instrucción sobre el cultivo del cacao en el estado de Tabasco." *Boletín de la Sociedad Mexicana de Geografía y Estadística* 2a época, vol. 4 (1872): 105–6.

Altamirano, Fernando. "Crónica de las excursiones científicas hechas por los miembros de la Sociedad." *Memorias y revista de la Sociedad Científica Antonio Alzate*, vol. 4 (1890): 123–30.

Altamirano, Fernando. "Necesidad de la repoblación de los bosques." Appendix to *La Naturaleza* series 2, vol. 2 (1897): 11–22.

Altamirano Piolle, María Elena. *José María Velasco: Paisajes de luz, horizontes de modernidad*. DGE Ediciones, 2006.

Altieri, Miguel A., and Victor Manuel Toledo. "The Agroecological Revolution in Latin America: Rescuing Nature, Ensuring Food Sovereignty and Empowering Peasants." *Journal of Peasant Studies* 38, no. 3 (2011): 587–612.

Álvarez Peláez, Raquel. *La conquista de la naturaleza americana*. CSIC, 1993.

Alzate y Ramírez, José Antonio de. "Botanica." Originally published in *Gaceta de literatura*, February 15, 1788. Republished in *Colección de documentos para la historical natural de México*, supplement to *La Naturaleza* vol. 4 (1884): 166–72.

Alzate y Ramírez, José Antonio de. *Gacetas de literatura*. Puebla, Mexico, 1831.

Alzate y Ramírez, José Antonio de. "Noticia de algunas plantas." Originally published in *Gaceta de literatura*, April 19, 1791. Republished in *Colección de documentos para la historia natural de México*, supplement to *La Naturaleza* vol. 7 (1887): 4–6.

Anderson, Benedict. *Imagined Communities: Reflections on the Origin and Spread of Nationalism*. Verso, 1991.

Arguilez, Ligia. "Making the Desert Place: An Environmental and Cultural History of the Creosote Bush from the US–Mexico Borderlands." PhD thesis, University of Texas–El Paso, 2023.

Arias Divito, Juan Carlos. *Las expediciones científicas españolas durante el siglo xviii*. Ediciones Cultura Hispanica, 1968.

Armus, Diego, and Pablo Gómez, eds. *The Gray Zones of Medicine: Healers and History in Latin America*. University of Pittsburgh Press, 2021.

Arriaga, José Joaquin. "Informe." Presented January 1873, *La Naturaleza* vol. 2 (1873): 262–74.

Arroyo, Miguel. "Que hace la Sociedad de Geografía y Estadística en el año de 1851." *Boletín de la Sociedad Mexicana de Geografía y Estadística* vol. 3 (1852): 3–5.

Azuela Bernal, Luz Fernanda. *Tres sociedades científicas en el porfirato*. UNAM, 1994.

Báez Macías, Eduardo. *Historia de la Escuela Nacional de Bellas Artes (Antigua Academia de San Carlos), 1781–1910*. UNAM, 2009.

Balbontín, Manuel. "El lago de Texcoco." *Boletín de la Sociedad Mexicana de Geografía y Estadística* 3a época, vol. 1 (1873): 371–77.

Balbontin, Manuel. "Los bosques." *Boletín de la Sociedad Mexicana de Geografía y Estadística* 3a época, vol. 1 (1873): 144–51.

Balbontín, Manuel. *Memorias del Coronel Manuel Balbontín*. Tip. de la Escuela I. Militar, 1896.

Bárcena, Mariano. "Aclimatación de plantas en la República." *La Naturaleza* vol. 2 (1873): 141–48.

Bárcena, Mariano. "Discurso al tomar posesión de la presidencia de la Sociedad." *La Naturaleza* vol. 5 (1882): 3–6.

Bárcena, Mariano. "El árbol de manitas." *La Naturaleza* vol. 3 (1876): 114–17.

Bárcena, Mariano. "El bosque de Chapultepec. Informe relative a las causas que origina le destrucción de su arboleda (1891)." *La Naturaleza* series 2a, vol. 2 (1897): 193–98.

Bárcena, Mariano. "El Jardín Botánico y de aclimatación de Guadalajara." *La Naturaleza* series 2, vol. 1 (1891): 433–36 and 437–42.

Bárcena, Mariano. "Informe." *La Naturaleza* vol. 3 (1876): 254–67.

Bárcena, Mariano. "La hauya elegans." *La Naturaleza* vol. 3 (1876): 305–7.

Bárcena, Mariano. "Observaciones de plantas características de climas y terrenos. Descripción de la bignonia viminalis." *La Naturaleza* vol. 2 (1873): 173–77.

Bárcena, Mariano, Miguel Pérez, Manuel Urbina, José Ramírez, and José Segura. "Dictámen sobre le repoblación vegetal del Valle de Mexico." *La Naturaleza* vol. 6 (1884): 245–51.

Baudot, George. *Utopia and History in Mexico: The First Chronicles of Mexican Civilization (1520–1569)*. Translated by Bernard R. Ortiz de Montellano and Thelma Ortiz de Montellano. University Press of Colorado, 1995.

Baver, Sherrie L. and Barbara Deutsch Lynch, eds. *Beyond Sun and Sand: Caribbean Environmentalisms*. Rutgers University Press, 2005.

Beltrán, Enrique. "Use and Conservation: Two Conflicting Principles." In *Proceedings of the First World Conference on National Parks International Union for the Conservation of Nature and Natural Resources, Seattle, Washington, 30 June–1 July 1962*, edited by Alexander B. Adams. National Park Service, U.S. Department of the Interior, 1964.

Benito-Vessels, Carmen. "Hernández in Mexico: Exile and Censorship?" In *Searching for the Secrets of Nature: The Life and Works of Dr. Francisco Hernández*, edited by Simon Varey, Rafael Chabrán, and Dora B. Weiner. Stanford University Press, 2000.

Blackbourn, David. *The Conquest of Nature: Water, Landscape, and the Making of Modern Germany*. W. W. Norton, 2006.

Bleichmar, Daniela. "Painting as Exploration: Visualizing Nature in Eighteenth-Century Colonial Science." *Colonial Latin American Review* 15, no. 1 (2006): 81–104.

Bleichmar, Daniela. "Training the Naturalist's Eye in the Eighteenth Century: Perfect Global Visions and Local Blind Spots." In *Visualizing the Unseen, Imagining the Un-*

known, Perfecting the Natural: Art and Science in the 18th and 19th Centuries, edited by Andrew Graciano. Cambridge Scholars Publishing, 2008.

Bleichmar, Daniela. *Visible Empire: Botanical Expeditions and Visual Culture in the Hispanic Enlightenment*. University of Chicago Press, 2012.

Bleichmar, Daniela, and Peter Mancall. "Introduction." In *Collecting Across Cultures: Material Exchanges in the Early Modern Atlantic World*, edited by Daniela Bleichmar and Peter Mancall. University of Pennsylvania Press, 2011.

Boone, Elizabeth Hill. *Descendants of Aztec Pictography: The Cultural Encyclopedias of Sixteenth-Century Mexico*. University of Texas Press, 2020.

Boone, Elizabeth Hill. "Maps of Territory, History and Community in Aztec Mexico." In *Cartographic Encounters: Perspectives on Native American Mapmaking and Map Use*, edited by G. Malcolm Lewis. University of Chicago Press, 1998.

Boone, Elizabeth Hill. "Migration Histories as Ritual Performance." In *To Change Place: Aztec Ceremonial Landscapes*, edited by David Carrasco. University of Colorado Press, 1991.

Borsay, Peter, with Rosemary Sweet. *The Invention of the English Landscape c. 1700–1939*. Bloomsbury, 2023.

Boruchoff, David A. "Anthropology, Reason, and the Dictates of Faith in the Antiquities of Francisco Hernández." In *Searching for the Secrets of Nature: The Life and Works of Dr. Francisco Hernández*, edited by Simon Varey, Rafael Chabrán, and Dora B. Weiner. Stanford University Press, 2000.

Boyer, Christopher R., ed. *A Land Between Waters: Environmental Histories of Modern Mexico*. University of Arizona Press, 2012.

Boyer, Christopher R. *Political Landscapes: Forests, Conservation, and Community in Mexico*. Duke University Press, 2015.

Brading, David. *The First America: The Spanish Monarchy, Creole Patriots, and the Liberal State, 1492–1867*. Cambridge University Press, 1991.

Brailovsky, Antonio Elio, and Dina Foguelman. *Memoria verde: Historia ecológica de la Argentina*. Editorial Sudamericana, 1991.

Bravo, Elba Mónica. "Abren banco de semillas en Xochimilco como estrategia ante el cambio climático." *La Jornada*, February 28, 2024.

Bray, David, and Peter Klepeis. "Deforestation, Forest Transitions, and Institutions for Sustainability in Southeastern Mexico, 1900–2000." *Environment and History* 11 (2005): 195–223.

Bray, David, Leticia Merino, and Deborah Barry, eds. *Los bosques comunitarios de México: Manejo sustentable de paisajes forestales*. Instituto Nacional de Ecología, 2007.

Brown, Donna. *Inventing New England: Regional Tourism in the Nineteenth Century*. Smithsonian, 1995.

Bullock, William. *Six Months Residence and Travels in Mexico*. London, 1824.

Burdet, Hervé. "Relato de Augustin Pyramus de Candolle sobre la elaboración de la *Flora de México*, también conocido como *Flora de las Damas de Ginebra*." In *El águila y el nopal: La expedición de Sessé y Mociño a Nueva España (1787–1803)*, edited by María Pilar de San Pío Aladrén and Miguel Angel Puig-Samper. Real Jardín Botánico, 2000.

Burke, Edmund. "A Philosophical Inquiry into the Origin of Our Ideas of the Sublime and Beautiful (1757)." In *Architectural Theory*, vol. 1., edited by H. F. Mallgrave. Blackwell, 2006.

Burkhart, Louise. *The Slippery Earth: Nahua-Christian Moral Dialogue in Sixteenth-Century Mexico*. University of Arizona Press, 1989.

Bustamante, Jesús. "The Natural History of New Spain." In *The Mexican Treasury: The Writings of Dr. Francisco Hernández*, edited by Simon Varey. Stanford University Press, 2000.

Bye, Robert. "The Role of Humans in the Diversification of Plants in Mexico." In *Biological Diversity of Mexico: Origins and Distribution*, edited by T. P. Ramamoorthy, Robert Bye, and Antonio Lot. Oxford University Press, 1993.

Candiani, Vera. *Dreaming of Dry Land: Environmental Transformation in Colonial Mexico City*. Stanford University Press, 2014.

Cañizares-Esguerra, Jorge. *Nature, Empire, and Nation: Explorations of the History of Science in the Iberian World*. Stanford University Press, 2006.

Cañizares-Esguerra, Jorge. "Postcolonialism *avante la lettre?* Travelers and Clerics in Eighteenth-Century Colonial Spanish America." In *After Spanish Rule: Postcolonial Predicaments of the Americas*, edited by Mark Thurner and Andrés Guerrero. Duke University Press, 2003.

Carey, Mark. *In the Shadow of Melting Glaciers: Climate Change and Andean Society*. Oxford University Press, 2010.

Carey, Mark. "Latin American Environmental History: Current Trends, Interdisciplinary Insights, and Future Directions." *Environmental History* 14 (2009): 221–52.

Carrasco, David, and Scott Sessions. *Daily Life of the Aztecs*, 2nd ed. Greenwood-ABC-CLIO, 2011.

Carrera, Magali M. *Traveling from New Spain to Mexico: Mapping Practices of Nineteenth-Century Mexico*. Duke University Press, 2011.

Carrillo, Crescencio. "Disertacion sobre la literature antigua de Yucatan." *Boletín de la Sociedad Mexicana de Geografía y Estadística* 2a época, vol. 3 (1871): 257–71.

Carruthers, David, ed. *Environmental Justice in Latin America: Problems, Promise, and Practice*. MIT Press, 2008.

Casado de Otaola, Santos. *Naturaleza patria: Ciencia y sentimiento de la naturaleza en la España del regeneracionismo*. Marcial Pons, 2010.

Casas, Carlos. "Botánica, raiz de jalapa." *Boletín de la Sociedad Mexicana de Geografía y Estadística* 2a época, vol. 1 (1869): 6–8.

Castañeda, Carlos Eduardo. "Introduction." In *Life in the Imperial and Loyal City of Mexico in New Spain and the Royal and Pontifical University of Mexico as Described in the Dialogues for the Study of the Latin Language Prepared by Francisco Cervantes de Salazar for Use in his Classes and Printed in 1554 by Juan Pablos*, by Francisco Cervantes de Salazar. Facsimile and translation by Minnie Lee Barrett Shepard. University of Texas Press, 1953.

Castro, Casimiro. *Mexico y sus alrededores: Colección de vistas monumentales, paisajes y trajes del país*. Mexico, 1869 [1855].

Castro, Casimiro, A. Signone, and Antonio García Cubas, *Álbum del ferrocarril mexicano: colección de vistas pintadas del natural*. Mexico, 1877.

Castro Herrera, Guillermo. *Los trabajos de ajuste y combate: Naturaleza y sociedad en la historia de América Latina*. Casa de las Américas, 1995.

"Catálogo de la colección de Drogas indígenas," *La Naturaleza* vol. 3 (1876): 424–25.

"Catálogo de la colección de productos naturales indígenas." *La Naturaleza* vol. 3 (1876): 382–92 and 399–418.

Celis de la Cruz, Martha. "Presencia de la masonería alemana en México: Carlos de Gagern (1826–1885)," *Revista de estudios históricos de la Masonería Latinoamerican y Caribeña* (Costa Rica) 2, no. 2 (2010–2011): 152–59.

Cervantes, Fernando. *The Devil in the New World: The Impact of Diabolism in New Spain*. Yale University Press, 1997.

Cervantes, Vicente. "La botánica moderna en el Jardín Botánico de Nueva España (1793)." In *Historia de la ciencia en México*, vol. 3, edited by Elías Trabulse. FCU, 1985.

Cervantes, Vicente. "Respuesta del discípulo a la carta satisfactoria en la que se impugnan las objeciones que propone su autor contra el sistema del Linneo (July 15, 1788)." In *Linneo en México: Las controversias sobre el sistema binario sexual, 1788–1798*, edited by Roberto Moreno. UNAM, 1989.

Cervantes de Salazar, Francisco. *Crónica de la Nueva España*, bk 1 (ca. 1565). Linkgua, 2010.

Chabrán, Rafeal. "The Classical Tradition in Renaissance Spain and New Trends in Philology, Medicine, and Materia Medica." In *Searching for the Secrets of Nature: The Life and Works of Dr. Francisco Hernández*, edited by Simon Varey, Rafael Chabrán, and Dora B. Weiner. Stanford University Press, 2000.

Chabrán, Rafeal, and Simon Varey. "Entr'acte." In *Searching for the Secrets of Nature: The Life and Works of Dr. Francisco Hernández*, edited by Simon Varey, Rafael Chabrán, and Dora B. Weiner. Stanford University Press, 2000.

Chabrán, Rafeal, and Simon Varey. "The Hernández Texts." In *The Mexican Treasury: The Writings of Dr. Francisco Hernández*, edited by Simon Varey. Stanford University Press, 2000.

Chabrán, Rafeal, and Simon Varey. "Medical Natural History in the Renaissance: The Strange Case of Francisco Hernández." *Huntington Library Quarterly* 57, no. 2 (1994): 124–51.

Charlot, Jean. *Mexican Art and the Academy of San Carlos, 1785–1915*. University of Texas Press, 1963.

Charmentier, Isabelle. "Carl Linnaeus and the Visual Representation of Nature." *Historical Studies in the Natural Sciences* 41, no. 4 (2011): 365–404.

Chomsky, Aviva, and Steve Striffler. "Empire, Labor, and Environment: Coal Mining and Anticapitalist Environmentalism in the Americas." *International Labor and Working-Class History* 85 (2014): 194–200.

Clavijero, Francisco Javier. "Breve noticia de las plantas y animales de México (1780)." *Colección de documentos para la historical natural de México*, supplement to *La Naturaleza* vol. 4 (1884): 5–60.

Clavijero, Francisco Javier. "Disertacion IV. Sobre los animals del reino de Mexico." Originally published in 1780. Republished in *Colección de documentos para la historical natural de México*, supplement to *La Naturaleza* vol. 6 (1884): 60–97.

Clavijero, Francisco Javier. *The History of Mexico, Volume 3*. Thomas Dobson, 1817.

Coatsworth, John. *Growth Against Development: The Economic Impact of Railroads in Porfirian Mexico*. Northern Illinois University, 1981.

Collinson, Helen, ed. *Green Guerillas: Environmental Conflicts and Initiatives in Latin America and the Caribbean*. Latin American Bureau, 1996.

Connor, Walker. "The Timelessness of Nations." In *History and National Destiny: Ethnosymbolism and Its Critics*, edited by Montserrat Guibernau and John Hutchinson. Blackwell, 2004.

Consentino, Delia, and Adriana Zavala, *Resurrecting Tenochtitlan: Imagining the Aztec Capital in Modern Mexico City*. University of Texas, 2023.

Cooley, Mackenzie. "Bezoar: Medicine in the Belly of the Beast." In *Natural Things in Early Modern Worlds*, edited by Mackenzie Cooley, Anna Toledano, and Duygu Yildirim. Routledge, 2023.

Cooley, Mackenzie. "The Giant Remains: Mesoamerican Natural History, Medicine, and the Cycles of Empire." *Isis* 112, no. 1 (2021): 45–67.

Cooley, Mackenzie. *The Perfection of Nature: Animals, Breeding, and Race in the Renaissance*. University of Chicago Press, 2022.

Cope, R. Douglas. *The Limits of Racial Domination: Plebeian Society in Colonial Mexico City, 1660–1720*. University of Wisconsin Press, 1994.

Cordero y Hoyos, Francisco. "Herbario de plantas indígenas." *La Naturaleza* vol. 3 (1876): 424–25.

Cortés, Hernán. "Second Letter." In *Letters from Mexico*, rev. ed., translated and edited by Anthony Pagden. Yale University Press, 1986.

Covarrubias, José María. "Apuntes estadísticos de la municipalidad de Tenamastlán en el estado de Jalisco." *Boletín de la Sociedad Mexicana de Geografía y Estadística* 2a época, vol. 3 (1871): 280–95.

Craib, Raymond. *Cartographic Mexico: A History of State Fixations and Fugitive Landscapes*. Duke University Press, 2004.

Cronon, William. "The Trouble with Wilderness; Or, Getting Back to the Wrong Nature." In *Uncommon Ground: Rethinking the Human Place in Nature*, edited by William Cronon. W. W. Norton, 1995.

Crosby, Alfred W. *The Columbian Exchange: Biological and Cultural Consequences of 1492*. Greenwood Publishing, 1972.

Cruz Soto, Rosalba. "Las publicaciones periódicas y la formación de una identidad nacional." *Estudios de historia moderna y contemporánea* (México) 20 (2000): 15–39.

"¿Cuál debe ser el árbol nacional?" *El Universal*, July 1, 1921.

Das, Pallavi V. *Colonialism, Development, and the Environment: Railways and Deforestation in British India, 1860–1884*. Palgrave Macmillan, 2015.

Davis, Diana K. *Resurrecting the Granary of Rome: Environmental History and French Colonial Expansion in North Africa*. Ohio University Press, 2007.

De Candolle, Augustin Pyramus. "Flora de México o colección de plantas raras o poco conocidas observadas en México y en la Nueva España." In *La Real Expedición Botánica a Nueva España*. Coordinated by Jaime Labastida. UNAM, 2010.

De Gortari, Elí. *La ciencia en la historia de México*. Editorial Grijalbo, 1980.

De Fleury, S. E. L. Carron. "Notas geológicas y estadísticas sobre Sonora y la Baja California." *Boletín de la Sociedad Mexicana de Geografía y Estadística* 2a época, vol. 1 (1869): 112–18.

de la Cadena, Marisol. *Indigenous Mestizos: The Politics of Race and Culture in Cuzco, Peru, 1919–1991*. Duke University Press, 2000.

de la Llave, Pablo. "Memoria acerca de una nueva especie de zapote." Originally published in *Registro Trimestre*, Mexico, February 6, 1832. Republished in 1884 in *Colección de documentos para la historia natural de México* by *La Naturaleza* supplement to vol. 7 (1887): 66–68.

De Vos, Paula. *Compound Remedies: Galenic Pharmacy from the Ancient Mediterranean to New World Spain*. University of Pittsburgh Press, 2020.

De Vos, Paula. "Methodological Challenges Involved in Compiling the Nahua Pharmacopeia." *History of Science* 55, no. 2 (2017): 210–33.

Dean, Carolyn, and Dana Leibsohn. "Hybridity and Its Discontents: Considering Visual Culture in Colonial Spanish America." *Colonial Latin American Review* 12, no. 1 (2003): 5–35.

Dean, Warren. *With Broadax and Firebrand: The Destruction of the Brazilian Atlantic Forest*. University of California Press, 1995.

Dedina, Serge. *Saving the Gray Whale: People, Politics, and Conservation in Baja California*. University of Arizona Press, 2000.

Dell'Amore, Christine. "Animals Stopped in Their Tracks by Border Wall?" *National Geographic News*, April 17, 2009. https://web.archive.org/web/20090423044155/http://news.nationalgeographic.com/news/2009/04/photogalleries/us-mexico-border-fence-wildlife/index.html?source=rss. Accessed April 30, 2009.

DeParle, Jason. "Next Generation of Conservative." *New York Times*, July 14, 2005.

"Descripción de la cochinilla mixteca, y de su cria y beneficio," *Boletín de la Sociedad Mexicana de Geografía y Estadística* vol. 3 (1852): 82–86.

Díaz, Lilia. "El Jardín Botánico de Nueva España y la obra de Sessé según documentos mexicanos," *Historia Mexicana* 27 (1977): 49–78.

Dietrich, Jakob (Jacobus Theodorus Tabernaemontanus). *Neuw Kreuterbuch, mit schoen, kunstlichen und leblichen Figuren unnd Conterfeyten, aller Gewachss der Kreuter*. Nicholaus Basseus, 1588–1591.

Downing, Theodore, Susanna B. Hecht, Henry A. Pearson, and Carmen Garcia-Downing, eds. *Development or Destruction: The Conversion of Tropical Forest to Pasture in Latin America*. Westview Press, 1992.

Drayton, Richard. *Nature's Government: Science, Imperial Britain and the 'Improvement' of the World*. Yale University Press, 2000.

Driscoll, John Paul. *All That is Glorious Around Us: Paintings from the Hudson River School*. Cornell University Press, 1997.

Dunlap, Thomas R. *Nature and the English Diaspora: Environment and History in the United States, Canada, Australia, and New Zealand*. Cambridge University Press, 1999.

Edwards, Clinton R. "The Human Impact on the Forest in Quintana Roo, Mexico." *Journal of Forest History* 30, no. 3 (1986): 120–27.

Emmart, Emily Walcott. "Introduction." In *The Badianus Manuscript: Codex Barberini (An Aztec Herbal of 1552)*. Facsimile with translation by Emily Walcott Emmart. Johns Hopkins Press, 1940.

Enciclopedia nacional popular de las actividades sociales, industriales, comerciales, y agrícolas de la República Mexicana. Mexico City, 1937.

Escobar, Sebastián. "Noticias geográficas y estadísticas del departamento de Soconusco [Estado de Chiapas]." *Boletín de la Sociedad Mexicana de Geografía y Estadística* 2a época, vol. 3 (1871): 76–92.

"Expedición al Virreinato de Nueva España, Tras las huellas de F. Hernández." *Real Jardín Botánico*. https://rjb.csic.es/expedicion-al-virreinato-de-nueva-espana-tras-las-huellas-de-f-hernandez-1787-1803/. Accessed March 17, 2008.

Ezcurra, Ezequiel. *De las chinampas a la megalópolis: El medio ambiente en la cuenca de México*. Fondo de Cultura Econeomica, 1990.

Fabricant, Nicole, and Kathryn Hicks. "Bolivia's Next Water War." *Radical History Review* no. 116 (2013): 130–45.

Fernández, Vicente. "Analysis del mezcal." *La Naturaleza* vol. 3 (1876): 363–64.

Fernández Caleya, Paloma Blanca. "Colecciones históricas: El herbario de la Real Expedición Botánica de Nueva España (1787–1803)." *Boletín de la Asociación de Herbarios Ibero-Macronénsicos* no. 5 (2000): 5–10.

Finck, Hugo. "Una excursión a las faldas del Pico de Orizava." *La Naturaleza* vol. 3 (1876): 231–35.

Fireman, Janet, and Manuel Servin. "Miguel Costanso: California's Forgotten Founder." *California Historical Society Quarterly* 49, no. 1 (1970): 3–19.

Flannery, Maura. *In the Herbarium: The Hidden World of Collecting and Preserving Plants*. Yale University Press, 2023.

Florescano, Enrique. *La bandera Mexicana: Breve historia de su formación y simbolismo*. Fondo de Cultura Económica, 2004.

Foster, Jeremy. *Washed with Sun: Landscape and the Making of White South Africa*. University of Pittsburg, 2008.

Freedberg, David. *The Eye of the Lynx*. University of Chicago Press, 2002.

Freitas, Frederico. *Nationalizing Nature: Iguazu Falls and National Parks at the Brazil–Argentina Border*. Cambridge University Press, 2021.

Fuentes, Carlos. *The Campaign*. Harper Perennial, 1992.

Funes Monzote, Reinaldo. *De bosque a sabana: Azúcar, deforestación y medio ambiente en Cuba, 1492–1926*. Siglo XXI, 2004.

Gagern, Carlos de. "Rasgos característicos de la raza indígena de México." *Boletín de la Sociedad Mexicana de Geografía y Estadística* 2a época, vol. 1 (1869): 802–18.

Gallini, Stefania. *Una historia ambiental del café en Guatemala: La Costa Cuca entre 1830 y 1902*. AVANCSO, 2009.

Garavaglia, Juan Carlos. "La pampa como ecosistema." In *Estudios sobre historia y ambiente en América, Volume 2: Norteamérica, Sudamérica y el Pacífico*, edited by Bernardo García Martínez and María del Rosario Prieto. El Colegio de México, 2002.

García, Crescencio. "Producciones utilísimas en los confines de los estados de Michoacán y Jalisco que pueden ser facilmente explotadas." *Boletín de la Sociedad Mexicana de Geografía y Estadística* 2a época, vol. 4 (1872): 557–69.

García Barragán, María Elisa. "El paisajista Adolfo Tenorio." *Anales del Instituto de Investigaciones Estéticas* 13: 193–203.

García Cubas, Antonio. *Atlas pintoresco é histórico de los Estados Unidos Mexicanos*. Mexico, 1885.

García Cubas, Antonio. "Reseña de los trabajos ejecutados por la Sociedad de Geografía y Estadística durante el ano de 1869." *Boletín de la Sociedad Mexicana Geografía y Estadística* 2a época, vol. 1 (1869): 932–43.

García Cubas, Antonio "Sociedad de Geografía y Estadística." *Boletín de la Sociedad Mexicana de Geografía y Estadística* 2a época, vol. 1 (1869): 1–2.

García Cubas, Antonio, Francisco Díaz Covarrubias, Manuel Fernández, and José Guadalupe Romero. "Dictamen sobre los inconvenientes de variar los nombres a los lugares de la República." *Boletín de la Sociedad Mexicana de Geografía y Estadística* 2a época, vol. 1 (1869): 601–4.

García Ramos, José. "Elogio histórico del farmacéutico Don Vicente Cervantes, Catedrático que fue de la Botánica." *Boletín de la Sociedad Mexicana de Geografía y Estadística*, series 2, vol. 1 (1869): 753–65.

Garrigan, Shelley. *Collecting Mexico: Museums, Monuments, and the Creation of National Identity*. University of Minnesota Press, 2012.

Gerbi, Antonello. *Nature in the New World: From Christopher Columbus to Gonzalo Fernández de Oviedo*, translated by Jeremy Moyle. University of Pittsburgh Press, 1986.

Gil, José Hilarión Romero. "Destrucción de los bosques en el Estado de Jalisco." *Boletín de la Sociedad Mexicana de Geografía y Estadística* 2a época, vol. 1 (1869): 10–14.

Gimmel, Millie. "Hacia una reconsideración del Códice de la Cruz Badiano: Nuevas propuestas para el studio de la medicina indígena en el period colonial." *Colonial Latin American Review* 17, no. 2 (2008): 273–83.

Godínez Ortega, José Luis. "Martín de Sessé y Lacasta, director de la Real Expedición." In *La Real Expedición Botánica a Nueva España*. Coordinated by Jaime Labastida. UNAM, 2010.

Golley, Frank Benjamin. *A History of the Ecosystem Concept in Ecology: More Than the Sum of the Parts*. Yale University Press, 1993.

González, Eleuterio. "Discurso sobre el Estudio de la Botánica." *La Naturaleza* vol. 5 (1882): 172–82.

González, Eleuterio. "La florula de la ciudad de Monterrey y sus inmediaciones." *La Naturaleza* vol. 3 (1876): 145–50.

Gonzalez, Miguel. "Algodón: Informe sobre el cultivo y producción del algodón en el Distrito de Jalapa." *Boletín de la Sociedad Mexicana de Geografía y Estadística* 2a época, vol. 1 (1869): 891–95.

González Bueno, Antonio. "Conocimiento científico y poder en la España Ilustrada." *Antilia, Revista Española de las ciencias de la Naturaleza y de la Tecnología* 1, no. 2 (1995): unpaginated.

González Bueno, Antonio, and Raúl Rodríguez Nozal, *Plantas americanas para la España ilustrada: Génesis, desarrollo y ocaso del projecto español de expediciones botánicas*. Editorial Complutense, 2000.

González Cos, Jesús. "Estadística del partido de Silao de la Victoria." *Boletín de la Sociedad Mexicana de Geografía y Estadística* 2a época, vol. 4 (1872): 301–16.

Grandia, Liza. *Enclosed: Conservation, Cattle, and Commerce Among the Q'eqchi' Maya Lowlanders*. University of Washington Press, 2012.

Granziera, Patrizia. "Huaxtepec: The Sacred Garden of an Aztec Emperor." *Landscape Research* 30, no. 1 (2005): 81–107.

Green, Nicholas. *The Spectacle of Nature: Landscape and Bourgeois Culture in Nineteenth-Century France*. Manchester University Press, 1990.

Guerrini, Luigi. *The Accademia dei Lincei and the New World*. Max Planck Institute, 2008.

Guy, Kolleen. *When Champagne Became French: Wine and the Making of a National Identity*. John Hopkins University Press, 2003.

Haber, Stephen H. *Industry and Underdevelopment: The Industrialization of Mexico, 1890–1940*. Stanford University Press, 1989.

Hale, Charles. *Liberalism in the Age of Mora, 1821–1853*. Yale University Press, 1968.

Hall, Stuart. "Gramsci's Relevance for the Study of Race and Ethnicity." In *Stuart Hall: Critical Dialogues in Cultural Studies*, edited by David Morley and Kuan-Hsing Chen. Routledge, 1996.

Hämäläinen, Pekka. "The Politics of Grass: European Expansion, Ecological Change, and Indigenous Power in the Southwest Borderlands." *The William and Mary Quarterly* 67, no. 2 (2010): 173–208.

Harper, Kyle. *Plagues upon the Earth: Disease and the Course of Human History*. Princeton University Press, 2021.

Hassig, Debra. "Transplanted Medicine: Colonial Mexican Herbals of the Sixteenth Century." *RES: Anthropology and Aesthetics* no. 17/18 (1989): 30–53.

Hernández, Francisco, Rafael Chabrán, and Simon Varey. "'An Epistle to Arias Montano': An English Translation of a poem by Francisco Hernández," *Huntington Library Quarterly* 55, no. 4 (1992): 621–34.

Hernández, Kelly Lytle. *Bad Mexicans: Race, Empire and Revolution in the Borderlands*. W. W. Norton, 2022.

Hernández Sáenz, Luz María. *Carving a Niche: The Medical Profession in Mexico, 1800–1879*. McGill–Queen's University Press, 2018.

Hernández y Dávalos, J. E. "Estado de Jalisco. Materiales para un diccionario geográfico, estadístico, historico y biografico del mismo. Municipio, villa, laguna e islas de Atoyac." *Boletín de la Sociedad Mexicana de Geografía y Estadística* 2a época, vol. 3 (1871): 177–203.

Herrera, Alfonso. "Apuntes para la materia médica mexicana. El yoyote." *La Naturaleza* vol. 2 (1873): 187–94.

Herrera, Alfonso. "Informe acerca de los trabajos de la Sociedad Mexicana de Historia Natural durante los años de 1890 y 1891." *La Naturaleza* 2a serie, vol. 2 (1897): 129–55.

Herrera, Alfonso. "Revista científica: El plátano." *La Naturaleza* supplement to vol. 5 (1882): 17–20.

Herrera, Alfonso. "Sinonimia vulgar y científica de algunas plantas silvestres y de varias de las que se cultivan en México." *La Naturaleza* vol. 2 (1873): 318–23 and 373–74.

Hill, Juan. "Cera vegetal." *Boletín de la Sociedad Mexicana de Geografía y Estadística* 2a época, vol. 1 (1869): 889–91.

Holguin, Sandie. *Creating Spaniards: Culture and National Identity in Republican Spain.* University of Wisconsin Press, 2002.

Humboldt, Alexander von. *Political Essay on the Kingdom of New Spain: Containing Researches Relative to the Geography of Mexico.* Longman, Hurst, Rees, Orme, and Brown, 1822.

Ibarra Cabrera, Silvio. "Profesor Don Miguel Bustamente y Septiem." In *Folletos de divulgación cientifica. contribuciones a la historia de las ciencias biológicas en México,* vol 2. Casa del Lago de Chapultepec, 1936.

Imort, Michael. "The Sylvan People: Wilhelmine Forestry and the Forest as a Symbol of Germandom." In *Germany's Nature: Cultural Landscapes and Environmental History,* edited by Thomas Lekan and Thomas Zeller. Rutgers University Press, 2005.

Instituto Médico Nacional, *Datos para la materia médica mexicana,* vol. 1. Secretaría de Fomento, 1894–1908.

"Introduction." *Boletín de la Sociedad Mexicana de Geografía y Estadística* 2a época, vol. 1 (1869): iii–vi.

Jiménez de Cisneros, Francisco. *Complutensian Polyglot Bible.* Spain, 1514.

Klooster, Dan. "Campesinos and Mexican Forest Policy During the Twentieth Century." *Latin American Research Review* 38, no. 2 (June 2003): 94–126.

Klubock, Thomas. *La Frontera: Forests and Ecological Conflict in Chile's Frontier Territory.* Duke University Press, 2014.

Kusukawa, Sachiko. *Picturing the Book of Nature: Image, Text, and Argument in Sixteenth-Century Human Anatomy and Medical Botany.* University of Chicago Press, 2012.

Labastida, Jaime. "Ilustración e independencia: José Mariano Mociño y la ciencia moderna." In *La Real Expedición Botánica a Nueva España.* Coordinated by Jaime Labastida. UNAM, 2010.

Labastida, Jaime, coordinator. *La Real Expedición Botánica a Nueva España.* UNAM, 2010.

Laird, Andrew. "Nahua Latinists: Classical Learning and Indigenous Legacies in Sixteenth-Century Mexico." In *Brill's Companion to Classics in the Early Americas,* edited by Matthew Duquès, Maya Feile Tomes, and Adam Goldwyn. Brill, 2021.

Larrainzar, Manuel. "Noticia histórica de Soconusco." *Boletín de la Sociedad Mexicana de Geografía y Estadística* vol. 3 (1852): 87–168.

Leal, Claudia. *Landscapes of Freedom: Building a Postemancipation Society in the Rainforests of Western Colombia.* University of Arizona Press, 2018.

Lekan, Thomas M. *Imagining the Nation in Nature: Landscape Preservation and German Identity, 1885–1945.* Harvard University Press, 2004.

León, Nicolás. *Botánico-Mexico: Catálogo, biográfico y crítico de autores y escritos referentes a vegetales de México y sus aplicaciones desdes la conquista hasta el presente.* Secretaría de Fomento, 1895.

Leopold, Aldo. *Sand Country Almanac.* Oxford University Press, 1948.

López, Rick A. *Crafting Mexico: Intellectuals, Artisans, and the State After the Revolution.* Duke University Press, 2010.

López Piñero, José M., and José Pardo Tomás. "The Contribution of Hernández to European Botany and Materia Medica." In *Searching for the Secrets of Nature: The Life and Works of Dr. Francisco Hernández,* edited by Simon Varey, Rafael Chabrán, and Dora B. Weiner. Stanford University Press, 2000.

López Piñero, José, and José Pardo Tomás, *Nuevos materiales y noticias sobre la Historia de las Plantas de Nueva España, de Francisco Hernández.* Universitàt de València, 1994.

López-Alonso, Moramay. *Measuring Up: A History of Living Standards in Mexico, 1850–1950.* Stanford University Press, 2012.

Lozoya, Xavier, ed. *Plantas y luces en México: La Expedición Científica a Nueva España (1787–1803).* Serbal, 1984.

Maehle, Andreas-Holger. *Drugs on Trial: Experimental Pharmacology and Therapeutic Innovation in the Eighteenth Century.* Rodopi Press, 1999.

Malanco, Luis. "Discurso pronunciado en la session del dia 31 de diciembre al tomar asiento por primera vez entre los miembros de la Sociedad." *Boletín de la Sociedad Mexicana de Geografía y Estadística* 2a época, vol. 1 (1869): 148–50.

Maldonado, Manuel. "La vida y la obra de Don Miguel Bustamante." *Revista de la Sociedad Mexicana de Historia Natural* 1, no. 3 (1940): 203–13.

Maldonado Polo, José Luis. *De California a El Petén: El naturalista riojano José Longinos Martínez en Nueva Espana.* Gobierno de La Rioja, 1997.

Maldonado Polo, José Luis. "El primer gabinete de historia natural de México y el reconocimiento del noroeste novohispano." *Estudios de historia novohispana* vol. 21 (2000): 49–66.

Maldonado Polo, José Luis. "La Expedición Botánica a Nueva España, 1786–1803: El Jardín Botánico y la Cátedra de Botánica." *Historia Mexicana* 50, no. 1 (2000): 5–56.

Marquardt, Steve. "'Green Havoc': Panama Disease, Environmental Change, and Labor Process in the Central American Banana Industry." *American Historical Review* 106, no. 1 (2001): 49–80.

Marroquín Arredondo, Jaime. "The Method of Francisco Hernández: Early Modern Science and the Translation of Mesoamerica's Natural History." In *Translating Nature: Cross-Cultural Histories of Early Modern Science,* edited by Jaime Marroquín Arredondo and Ralph Bauer. University of Pennsylvania Press, 2019.

Martínez-Altieri, Joan. "Ecology and the Poor: A Neglected Dimension of Latin American History." *Journal of Latin American Studies* 23, no. 3 (1991): 621–39.

Martínez-Hernández, Gerardo. "La llegada del cirujano Alonso López de Hinojosos a la Nueva España." *Revista médica del Instituto Mexicano de Seguro Social* 49, no. 4 (2021): 459–63.

Martins, Carlos. "Las poblaciónes vegetales. Su origen, su composión y sus emigraciones." *La Naturaleza* vol. 2 (1873): 148–54 and 241–50.

Matchett, Karin. "At Odds over Inbreeding: An Abandoned Attempt at Mexico/United States Collaboration to 'Improve' Mexican Corn, 1940–1950." *Journal of the History of Biology* 39 (2006): 345–71.

McCook, Stuart. *States of Nature: Science, Agriculture, and Environment in the Spanish Caribbean, 1760–1940*. University of Texas Press, 2002.

McDonough, Kelly S. *Indigenous Science and Technology: Nahuas and the World Around Them*. University of Arizona Press, 2024.

McManus, Stuart. "Decolonizing Renaissance Humanism." *American Historical Review* 127, no. 3 (2022): 1131–61.

McVaugh, Rogers. *Botanical Results of the Sessé and Mociño Expedition (1787–1803)*. Carnegie Mellon University Press, 2000.

Melillo, Edward. *Strangers on Familiar Soil: Rediscovering the Chile-California Connection*. Yale University Press, 2015.

Melville, Elinor G. K. *A Plague of Sheep: Environmental Consequences of the Conquest of Mexico*. Cambridge University Press, 1994.

Méndez Reyes, Salvador. "Instituciones, obras y viajeros." In *La formación geográfica de México*, edited by Carlos Herrejón Peredo. CONACULTA, 2011.

Mendiondo, Pedro. Introduction to "El Té." *Boletín de la Sociedad Mexicana de Geografía y Estadística* 2a época, vol. 2 (1870): 525–29.

Mendoza, Eufemio. "Discurso." *Boletín de la Sociedad Mexicana de Geografía y Estadística* 2a época, vol. 3 (1871): 204–8.

Mendoza, Gumesindo, and Alfonso Herrera. "El anacahuite." *La Naturaleza* vol. 3 (1876): 151–54.

"Mexican President Confident Congress Will Ban Planting GM Corn in 2025." *Reuters*, December 21, 2024. https://www.reuters.com/markets/commodities/mexican-president-confident-congress-will-ban-planting-gm-corn-2025-2024-12-21/. Accessed January 31, 2025.

Miller, Ian Jared, Julia Adeney Thomas, and Brett L. Walker, eds. *Japan at Nature's Edge: The Environmental Context of Global Power*. University of Hawai'i Press, 2013.

Miller, Perry. *Nature's Nation*. Harvard University Press, 1967.

Miller, Shawn. *An Environmental History of Latin America*. Cambridge University Press, 2007.

Miller, Shawn. "Stilt Root Subsistence: Colonial Mangroves and Brazil's Landless Poor." *Hispanic American Historical Review* 83, no. 2 (2003): 223–53.

Mociño, José Mariano. "Descripción del volcán de Tuxtla (1787)." *La Naturaleza* vol. 3 (1876): 106–14.

Mociño, José Mariano. "Discurso que dio en la apertura de las lecciones de botánica en México." Originally published in *Anales de Ciencias Naturales* vol. 5 (1802). Republished in *Colección de documentos para la historia natural de México*, a supplement to *La Naturaleza* vol. 7 (1887): 39–42.

Moncada Maya, José Omar. *El ingeniero Miguel Constanzó. Un military ilustrado en la Nueva España del siglo XVIII*. UNAM, 1994.

Moncada Maya, José Omar, and Patricia Gómez Rey. "Patrimonio geográfico mexicano del siglo XX." In *La formación geográfica de México*, edited by Carlos Herrejón Peredo. CONACULTA, 2011.

Morales, José D. "El ypacihuitl o la yerba de zorillo." La Naturaleza vol. 2 (1873): 194–98.

Morales Moreno, Luis Gerardo. *Orígenes de la museología mexicana: Fuentes para el estudio histórico del Museo Nacional, 1780–1940*. Universidad Iberoamericana, 1994.

Moreno, Roberto, ed. *Linneo en México: Las controversias sobre el sistema binario sexual, 1788–1798*. UNAM, 1989.

Morin González, Aidé. "Catecismo testeriano: Una lectura de evangelización." In *Memoria XVIII Encuentro Nacional de Investigadores del Pensamiento Novohispano*. Universidad de San Luis Potosí, 2005.

"Mountain Lion Stuck at Border Wall." *Demarcated Landscape: Conservation Geography*, September 19, 2008. https://demarcatedlandscapes.wordpress.com/2008/09/19/mountain-lion-stuck-at-border-wall/. Accessed April 30, 2009.

Mukerji, Chandra. *Territorial Ambitions and the Gardens of Versailles*. Cambridge University Press, 1997.

Myers, Kathleen Ann. *Fernández de Oviedo's Chronicle of America: A New History for a New World*. University of Texas Press, 2007.

Nash, Frederick Roderick. *Wilderness and the American Mind*. Yale University Press, 2012.

Neri, Janice. *The Insect and the Image: Visualizing Nature in Early Modern Europe, 1500–1700*. University of Minnesota, 2011.

Nickelson, Kärin. "Draughtsmen, Botanists and Nature: Constructing Eighteenth-Century Botanical Illustrations." *Studies in History and Philosophy of Biological and Biomedical Sciences* 37 (2006): 1–25.

Nieremberg, Juan Eusebio. *Historia naturae maxime peregrinae*. Antwerp, 1635.

Nieremberg, Juan Eusebio. "Selections from Juan Eusebio Nieremberg's transcriptions of Hernández, in *Historia naturae, maxime peregrine* (1635)." In *The Mexican Treasury: The Writings of Dr. Francisco Hernández*, edited by Simon Varey. Stanford University Press, 2000.

Nieto, Mauricio. "Presentación gráfica, desplazamiento y aprobación de la naturaleza en las expediciones botánicas del siglo XVIII." *Asclepio* 47, no. 2 (1995): 91–107.

Noriega, Tomás. "El ahuehuete." *La Naturaleza* vol. 4 (1879 [1880]): 35–40.

Nuttall, Zelia. *The Gardens of Ancient Mexico*. Editorial Cultura México, 1937.

O'Brien, Karen. *Sacrificing the Forest: Environmental and Social Struggles in Chiapas*. Westview Press, 1998.

Oelschlaeger, Max. *The Idea of Wilderness*. Yale University Press, 1991.

Ogilvie, Brian. *The Science of Describing: Natural History in Renaissance Europe*. University of Chicago Press, 2006.

Olavide, Miguel Fuertes, Carlos Martínez Shaw, and Miguel Ángel Puig-Samper Mulero. "José Mariano Mociño: El destino de su obra y las cartas de De Candolle." *Espacio, Tiempo y Forma. Serie IV, Historia Moderna* 12 (1999): 443–77.

Oles, James. *Art and Architecture in Mexico*. Thames and Hudson, 2013.

Ortega Reyes, Manuel. "El gigante de la flora mexicana, el sabino de Santa María del tule." *La Naturaleza* vol. 6 (1884): 100–115.

Ortiz de Montellano, Bernardo. "El conocimiento de la naturaleza entre los Mexicanos, Taxonomía." In *Historia general de la medicina en México. Tomo I: México Antiguo*, coordinated by Alfredo López Austin and Carlso Viesca Treviño. UNAM, 1984.

Ortiz Monasterio, Fernando, Isabel Fernández, Alicia Castillo, José Ortiz Monasterio, and Alfonso Bulle Goyri, *Tierra profanada: Historia profanada de México.* INAH, 1987.

Pani, Erika. "Los viajeros decimonónicos y la definición de lo nuestro." In *La idea de nuestro patrimonio histórico y cultural,* edited by Pablo Escalante Gonzalbo. CONACULTRA, 2011.

Pasquel, Leonardo. *Estampas de la Ciudad de México: La Alameda Central.* Editorial Chapultepec, 1980.

Pavord, Anna. *The Naming of Names: The Search for Order in the World of Plants.* Bloomsbury, 2005.

Payno, Manuel. "Anáhuac-México-Tenochtitlan-América." *Boletín de la Sociedad Mexicana de Geografía y Estadística* 2a época, vol. 2 (1870): 204–8.

Payno, Manuel. "Bosques y arbolados." *Boletín de la Sociedad Mexicana de Geografía y Estadística* 2a época, vol. 2 (1870): 77–91.

Payno, Manuel. "El Maguey Mexicano y sus diversos productos." *Boletín de la Sociedad Mexicana de Geografía y Estadística* 10, no. 5 (1864): 283–88.

Payno, Manuel. "Introduction." *Boletín de la Sociedad Mexicana de Geografía y Estadística* 2a época, vol. 2 (1870): iii–iv.

Payno, Manuel. "Notas geológicas y estadísticas sobre Sonora y la Baja-California submitted Dec 1868," *Boletín de la Sociedad Mexicana de Geografía y Estadística* 2a época, vol. 1 (1869): 44–52.

Peñafiel, Antonio. "Apuntes de viaje. Paleontologia, botanica, zoologica." *La Naturaleza* vol. 2 (1873): 255–62.

Peñafiel, Antonio. *La Naturaleza* addendum at the end of issue 22, vol. 2 (1873), unpaginated.

Phaf-Rheinberger, Ineke. "Between History of Art and History of Science: A New Appraisal of José María Velasco." In *Handbook of the Historiography of Latin American Studies on the Life Sciences and Medicine,* edited by Ana Barahona. Springer, 2022.

Pimentel, Francisco. *Memoria sobre las causas que han originado la situación actual de la raza indígena de México y medios de remediarla.* Mexico City, 1864.

Pimentel, Francisco. "Observaciones de Francisco Pimentel a la disertación sobre el idioma Otomí." *Boletín de la Sociedad Mexicana de Geografía y Estadística* 2a época, vol. 4 (1872): 224–36.

Pimentel, Juan. "Baroque Natures: Juan E. Nieremberg, American Wonders, and the Preterimperial Natural History." In *Science in the Spanish and Portuguese Empires, 1500–1800,* edited by Daniela Bleichmar, Paula De Vos, Kristin Huffine, and Kevin Sheehan. Stanford University Press, 2008.

Plantas de Nueva España coleccionadas por mandato y bajo los auspicious del benignisimo Rey Carlos IV, y ordenadas conforme al system de Linneo. Sociedad Mexicana de Historia Natural/*La Naturaleza* 1887.

Potts, Juan. "Chihuahua. Apuntes descriptivos de ese estado." *Boletín de la Sociedad Mexicana de Geografía y Estadística* 2a época, vol. 4 (1872): 279–83.

Pratt, Mary Louise. *Imperial Eyes: Travel Writing and Transculturation.* Routledge, 1992.

"Prospecto." *La Naturaleza* vol. 4 (1879[1880]), unpaginated back matter.

Puga, Guillermo, Beltrán y. "El Cerro de la Estrella o de Ixtapalapa." *La Naturaleza* 2a serie, vol. 1 (1891): 488–92.

Puig-Samper, Miguel Ángel. "Illustrators of the New World: The Image in the Spanish Scientific Expeditions of the Enlightenment." *Culture and History Digital Journal* 1, no. 2 (2012): 1–28.

Puig-Samper, Miguel Ángel, and Sandra Rebok. "El reconocimiento oficial de Alexander von Humboldt en España." *Humboldt im Netz: Internacional Review for Humboldtian Studies* 5, no. 5 (2004): 1–13.

Pumian, Abraham. "Noticia geográfica y stadística del pueblo y municipio de Quechula" *Boletín de la Sociedad Mexicana de Geografía y Estadística* 2a época, vol. 4 (1872): 481–504.

Radding, Cynthia. "The Children of Mayahuel: Agaves, Human Cultures, and Desert Landscapes in Northern Mexico." *Environmental History* 17, no. 1 (2012): 84–115.

Radkau, Joachim. "Germany as a Focus of European 'Particularities' in Environmental History." In *Germany's Nature: Cultural Landscapes and Environmental History*, edited by Thomas Lekan and Thomas Zeller. Rutgers University Press, 2005.

Ramamoorthy, T. P., Robert Bye, Antonio Lot, and John Fa. "Introduction." In *Biological Diversity of Mexico: Origins and Distribution*, edited by T. P. Ramamoorthy, Robert Bye, Antonio Lot, and John Fa. Oxford University Press, 1993.

Ramírez, Ignacio. "Lecturas de historia política de México. Las naciones primitivas." *Boletín de la Sociedad Mexicana de Geografía y Estadística* 2a época, vol. 3 (1871): 68–75.

Ramírez, Ignacio, Gumesindo Mendoza, Luis Malanco, and Ignacio Cornejo. "Bosques y arbolados." Dated January 27, 1870. *Boletín de la Sociedad Mexicana de Geografía y Estadística* 2a época, vol. 2 (1870): 14–24.

Ramírez, Ricardo. "Introduction." In *Flora Mexicana*, 2nd ed., by Martín Sessé y Lacasta and Joseph Mariano Mociño. Secretaría de Fomento, 1894.

Ramírez, Ricardo. Introduction to "Reseña de la Expedición de Historia Natural dirigido por Martín Sessé." In *Flora Mexicana*, supplement issue to *La Naturaleza* series 2, vol. 2 (1897).

Reeds, Karen Meier. "Renaissance Humanism and Botany." *Annals of Science* 33 (1976): 519–42.

Revkin, Andrew. *The Burning Season: The Murder of Chico Mendez and the Fight for the Amazon Rain Forest*. Island Press, 1990.

Reyes, V. "Ligeras nociones acerca del Tepehuaje." *Boletín de la Sociedad de Geografía y Estadística de la República Mexicana* 2a época, vol. 4 (1872): 115–19.

Rickett, Harold William. "The Royal Botanical Expedition to New Spain (1788–1820) as described in documents at the AGN (Mexico)." *Cronica Botanica* vol. 2 (1947): 1–86.

Risse, Guenter B. "Shelter and Care for Natives and Colonists: Hospitals in Sixteenth-Century New Spain." *Searching for the Secrets of Nature: The Life and Works of Dr. Francisco Hernández*. Edited by Simon Varey, Rafael Chabrán, and Dora B. Weiner. Stanford University Press, 2000.

Riva Palacio, Vicente, ed. *México a través de los siglos*. Mexico, 1884.

Rivera y Mendoza, A. de "Informe sobre el cultivo y producción del café en el distrito de Jalapa." *Boletín de la Sociedad de Geografía y Estadística de la República Mexicana* 2a época, vol. 2 (1870): 212–17.

Robertson, Donald. *Mexican Manuscript Painting of the Early Colonial Period: The Metropolitan Schools*. University of Oklahoma Press, 1994.

Rodríguez, Paul. *Enlightened Immunity: Mexico's Experiments with Disease Prevention in the Age of Reason*. Stanford University Press, 2018.

Rodríquez Figueroa, Andrea B. *Los jardines nahuas prehispanicos*. UNAM, 2021.

Rodríguez Nozal, Raúl. "La Oficina Botánica (1788–1835): Una institución dedicada al estudio de la flora americana." *Asclepio* 47, no. 2 (1995): 169–83.

Romero, José Guadalupe. "Reseña de los trabajos científicos de la Sociedad Mexicana de Geografía y Estadística en el año de 1863." *Boletín de la Sociedad Mexicana de Geografía y Estadística* 10 (1863/1864): 71–76.

Romero-Huesca, Andrés, and Julio Ramírez Bolla. "La atención médica en el Hospital Real de Naturales." *Cirugía y cirujanos* 7, no. 6 (2003): 496–505.

Rousseau, Jean Jacques. *Letters on the Elements of Botany*. Printed for B. White and Son, 1785.

Rovirosa, José. "Recuerdos de la ascension a la Montaña de Loma-de-Caballo." *La Naturaleza* vol. 7 (1887): 270–84.

Ruiz, Eduardo. "Riqueza de Michoacán." *Boletín de la Sociedad de Geografía y Estadística de la República Mexicana* 2a época, vol. 4 (1872): 321–26.

Rutsch, Mechthild. "Natural History, National Museum and Anthropology in Mexico: Some Reference Points in the Forging and Re-Forging of National Identity." *Perspectivas Latinoamericanas* no. 1 (2004): 89–122.

Safford, W. E. "*Lignum Niephriticum*: Its History and an Account of the Remarkable Fluorescence of Its Infusion." In *Annual Report of the Board of Regents of The Smithsonian Institution, 1915*. Smithsonian, 1916.

Salonio, Antonio María. "Ordenanzas antiguas sobre bosques y necesidad de observarlas." *Boletín de la Sociedad Mexicana de Geografía y Estadística* 2a época, vol. 1 (1869): 15–20.

San Pío Aladrén, María Pilar de, and Miguel Ángel Puig-Samper, eds. *El águila y el nopal: La Expedición de Sessé y Mociño a Nueva España (1787–1803)*. Real Jardín Botánico, 2000.

Sánchez, Jesús. "Fundación del Museo Nacional de Historia Natural." *La Naturaleza* 3a serie, vol. 1 (1910–12): 1–6.

Sánchez de Lorenzo Cáceres, José Manuel. "Las plantas de la flora mexicana cultivadas en España." www.arbolesornamentales.es/Plantasmexicanas.pdf. Accessed January 5, 2024.

Sánchez Hernández, Sergio. "Instituto de Ciencias, Literatura y Artes." *Boletín del Archivo General de la Nación* 4, no. 2 (1994): 74–76.

Santiago, Myrna. *The Ecology of Oil: Environment, Labor, and Mexican Revolution, 1900–1938*. Cambridge University Press, 2009.

Sarabia, Angélica Morales. *La consolidación de la botánica mexicana: Un viaje por la obra del naturalista José Ramírez (1852–1904)*. UNAM, 2015.

Sarabia, Angélica Morales and Patricia Aceves Pastrana. "Datos para la materia médica (1894–1908): Plantas medicinales, terapéutica y nacionalismo." *Circumscribere: International Journal for the History of Science* 9 (2011): 11–28.

Sarmiento, F. J. Puerto, and A. González Bueno. "Política científica y expediciones botánicas en el program colonial español ilustrado." In *Mundialización de la ciencia y cultura nacional*, edited by A. Lafuente, A. Elena, and M. L. Ortega. Doce Calles, 1993.

Schama, Simon. *Landscape and Memory*. Vintage, 1996.

Schiebinger, Londa. *Plants and Empire: Colonial Bioprospecting in the Atlantic World*. Harvard University Press, 2007.

Schifter Aceves, Liliana, and Angélica Morales Sarabia. "La trayectoria de Francisco Río de la Loza en la Sección de Química Analítica del Instituto Médico Nacional." *Revista mexicana de ciencias farmacéuticas* 43, no. 4 (2012): 69–78.

Schmidt, Arthur. "The Social and Economic Effect of the Railroad in Puebla and Veracruz, 1867–1911." PhD thesis, Indiana University, 1974.

Schwartz, Stuart. *Sea of Storms: A History of Hurricanes in the Greater Caribbean from Columbus to Katrina*. Princeton University Press, 2015.

Schwarz, Astrid, and Kurt Jax, *Ecology Revisted: Reflecting on Concepts, Advancing Science*. Springer, 2011.

"Se anima el concurso para elegir el árbol nacional." *El Demócrata*, July 1, 1921.

Sedrez, Lise. "The 'Bay of All Beauties': State and Environment in Guanabara Bay, Rio de Janeiro, 1875–1975." PhD thesis, Stanford University, 2005.

"Selvicultura." *Boletín de la Sociedad Mexicana de Geografía y Estadística* 2a época, vol. 1 (1869): 9–20.

Sessé y Lacasta, Martín, and José María Mociño. "De las plantas y de las flores de la Nueva España." In *Historia de la ciencia en México*, vol. 3, edited by Elías Trabulse. FCU, 1985.

Sessé y Lacasta, Martín, and José María Mociño. *Flora Mexicana*. Mexico, 1887.

Sessé y Lacasta, Martín, and José María Mociño. *Plantae Novae Hispaniae*. Mexico, 1887.

Sessé y Lacasta, Martín, and José María Mociño. *Plantae Novae Hispaniae*, 2nd ed. Mexico, 1893.

Sherman, Michael, and William Meyer, *The Course of Mexican History*. Oxford University Press, 1991.

Siméon, Rémi. *Diccionario de la lengua Náhuatl o Mexicano*. Siglo XX, 2004.

Simon, Joel. *Endangered Mexico: An Environment on the Edge*. Sierra Book Club, 1997.

Simonian, Lane. *Defending the Land of the Jaguar: A History of Conservation in Mexico*. University of Texas Press, 1995.

Smardon, Richard, and Betty Faust. "International Policy in the Biosphere Reserves of Mexico's Yucatan Peninsula." *Landscape and Urban Planning* 74 (2006): 160–92.

Sociedad Mexicana de Geografía y Estadística, *Biografía del Señor don Benigno Bustamante y Septien*. Mexico, 1859.

Soluri, John. *Banana Cultures: Agriculture, Consumption, and Environmental Change in Honduras and the United States*. University of Texas Press, 2005.

Soluri, John, Claudia Leal, and José Pádua. "Introduction. Finding the 'Latin American' in Latin American Environmental History." In *A Living Past: Environmental Histories of Modern Latin America*, edited by John Soluri, Claudia Leal, and José Pádua. Berghahn Books, 2018.

Somolinos d'Ardois, Germán. *La primera expedición científica en América*. SEP, 1971.

Sosa, Francisco. "Biografias de naturalists mexicanos. Don José Mariano Mociño." *La Naturaleza* vol. 7 (1887): 113–16.

Sosa, Francisco. "Biografias de naturalists mexicanos. Don Miguel Bustamante." *La Naturaleza* vol. 7 (1887): 126–27.

Sosa-Escalante, Javier Enrique, and Fernando Álvarez. "El estudio de la biodiversidad en México: ¿Una ruta con dirección?" *Revista Mexicana de Biodiversidad*, suppl. 85 (2014): 1–9.

Soto Laveaga, Gabriela. *Jungle Laboratories: Mexican Peasants, National Projects, and the Making of the Pill*. Duke University Press, 2009.

Sowerby, James. *A Botanical Drawing-Book*. Richard Taylor and Co. [1788].

Steinberg, Ted. *Down to Earth: Nature's Role in American History*. Oxford University Press, 2008.

Stepan, Nancy. *Picturing Tropical Nature*. Cornell University Press, 2001.

Stott, Philip. "Jungles of the Mind: The Invention of the 'Tropical Rain Forest.'" *History Today* 51, no. 5 (2001): 38–44.

Studnicki-Gizbert, Daviken. *The Three Deaths of Cerro de San Pedro: Four Centuries of Extractivism in a Small Mexican Mining Town*. University of North Carolina Press, 2022.

Sutherland, Cirilo Nelson. "Material tipo de La Colección de Sessé y Mociño en el Real Jardín Botánico de Madrid." *Anales del Jardín Botánico de Madrid* 55, no. 2 (1997): 375–418.

Tanck de Estrada, Dorothy. "Aspectos de la intervención de Carlos III en la Universidad de México." *Historia Mexicana* 38, no. 2 (1988): 181–98.

Tenorio-Trillo, Mauricio. *Mexico at the World's Fairs: Crafting a Modern Nation*. University of California Press, 2018.

Tezozómoc, Fernando Alvarado. *Crónica Mexicana (1598)*. Edited by Manuel Orozco y Berra. Mexico, 1878.

Thomas, Julia Adeney. *Reconfiguring Modernity: Concepts of Nature in Japanese Political Ideology*. University of California Press, 2008.

Thomas, Keith. *Man and the Natural World: Changing Attitudes in England, 1500–1800*. Oxford University Press, 1983.

Thoreau, Henry David. "Walking." *The Atlantic Monthly: A Magazine of Literature, Art, and Politics* 9, no. 56 (1862): 657–74.

Tobin, Beth Fowkes. "Imperial Designs: Botanical Illustration and the British Botanical Empire." *Studies in Eighteenth-Century Culture* 25 (1996): 265–92.

Tobin, Beth Fowkes. *Picturing Imperial Power: Colonial Subjects in Eighteenth-Century British Painting*. Duke University Press, 1999.

Torres, Mauricio. "Sheinbaum formaliza su reforma para prohibir la siembra de maíz transgénico en México." *CNN en Español*, January 24, 2025. https://cnnespanol.cnn

.com/2025/01/24/mexico/sheinbaum-reforma-prohibir-maiz-transgenico-orix. Accessed January 31, 2025.

Torres Montero, María Gabriela. "Nature and Culture in Nineteenth-Century Mexico: The Sociedad Mexicana de Historia Natural (1868–1914)." PhD thesis, University of Kansas, 2014.

Tortolero Villaseñor, Alejandro, ed. *Tierra, agua y bosques: Historia y medio ambiente en el México central.* Universidad de Guadalajara, 1996.

Townsend, Camilla. *Fifth Sun: A New History of the Aztecs.* Oxford University Press, 2019.

Treviño Urquijo, María Cristina. *Comisión Geográfica-Exploradora, 1878–1914.* Mapoteca Orozco y Berra, 2017.

Tucker, Richard. *Insatiable Appetite: The United States and the Ecological Degradation of the Tropical World.* Rowman and Littlefield, 2007.

Turner, Frederick Jackson. "The Significance of the Frontier in American History." In *Rereading Frederick Jackson Turner: The Significance of the Frontier in American History,* with commentary by John Mack Faragher. Yale University Press, 1994.

UNESCO. "Task Force on Criteria and Guidelines for the Choice and Establishment of Biosphere Reserves." *MAB Report Series.* Report no. 22, 1974.

Varey, Simon. "Francisco Hernández, Renaissance Man." In *Searching for the Secrets of Nature: The Life and Works of Dr. Francisco Hernández,* edited by Simon Varey, Rafael Chabrán, and Dora B. Weiner. Stanford University Press, 2000.

Varey, Simon, ed. *The Mexican Treasury: The Writings of Dr. Francisco Hernández.* Translations by Rafael Chabrán, Cynthia L. Chamberlin, and Simon Varey. Stanford University Press, 2000.

Vega y Ortega, Rodrigo. "El desarrollo del jardín botánico a través de la prensa de la ciudad de México, 1832–1846." *Trashumante. Revista Americana de Historia Social,* 7 (2016): 52–72.

Vega y Ortega, Rodrigo. "'Una ciencia tan útil como agradable': Los discursos del catedrático Pío Bustamante y Rocha: Ilustración y romanticismo en la enseñanza botánica, 1845–1860." *Revista de El Colegio de San Luis* nueva época, 9, no. 18 (2019): 235–57.

Velasco, José María. "Discurso al dejar la vicepresidencia de la Sociedad." *La Naturaleza* 6 (1884): 4–9.

Villada, Manuel. "Apuntes de geología y de botánica relativos a México, erupciones de agua caliente en el estado de Michoacán." *La Naturaleza* 2a serie, vol. 1 (1891): 419–33.

Villada, Manuel. "Breve noticia de un viaje de exploración a diversos lugares del estado de Veracruz (1893)." *Anales del Museo Nacional de México* 2a época, vol. 4 (c. 1907): 553–76.

Villada, Manuel. "Crónica." *La Naturaleza* vol. 4 (1879 [1880]): 8.

Villada, Manuel. "Obituary of Manuel Urbina y Altamirano." *La Naturaleza* 3a serie, vol. 1, (1910–12): unpaginated.

Villada, Manuel, and Eduardo Armendariz. "Necesidad de la conservación de los bosques." Supplement to *La Naturaleza* 2a serie, vol. 2 (1897): 3–10.

Villar, Luis, ed. *Homenaje a Martín de Sessé y Juan del Castillo, naturalistas jacetonos del siglo XVIII.* Instituto de Estudios Altoaragoneses, 1993.

Vitale, Luis. *Hacia una historia del ambiente en América Latina: De las culturas aborígenes a la crisis ecológica actual.* Nueva Imagen, 1983.

Vivanco, Luis. *Green Encounters: Shaping and Contesting Environmentalism in Rural Costa Rica.* Berghahn Books, 2007.

Wakild, Emily. "Environmental Justice, Environmentalism, and Environmental History in Twentieth-Century Latin America." *History Compass* 11, no. 2 (2013): 163–76.

Wakild, Emily. "Parables of Chapultepec: Urban Parks, National Landscapes, and Contradictory Conservation in Modern Mexico." In *A Land Between Waters: Environmental Histories of Modern Mexico*, edited by Christopher R. Boyer. University of Arizona Press, 2012.

Wakild, Emily. *Revolutionary Parks: Conservation, Social Justice, and Mexico's National Parks, 1910–1940.* University of Arizona Press, 2011.

Walker, Charles. *Shaky Colonialism: The 1746 Earthquake-Tsunami in Lima, Peru, and Its Long Aftermath.* Duke University Press, 2008.

Webb, James Josiah. *Adventures in the Santa Fé Trade, 1844–1847.* Bison Books, 1995.

Weber, Eugen. *Peasants into Frenchmen: The Modernization of Rural France, 1870–1914.* Stanford University Press, 1976.

Weiner, Dora B. "The World of Dr. Francisco Hernández." In *Searching for the Secrets of Nature: The Life and Works of Dr. Francisco Hernández.* Edited by Simon Varey, Rafael Chabrán, and Dora B. Weiner. Stanford University Press, 2000.

Weiner, Dora B., and Katharine E. Donahue. "The Doctors Simon Varey and Francisco Hernández at UCLA: Tragedy and Triumph." In *Sustaining Literature: Essays on Literature, History, and Culture, 1500–1800*, edited by Greg Clingham. Bucknell University Press, 2007.

Whited, Tamara. *Forests and Peasant Politics in Modern France.* Yale University Press, 2000.

Widdifield, Stacie G. *The Embodiment of the National in Latin Nineteenth-Century Mexican Painting.* University of Arizona Press, 1996.

Williams, Barbara J. "Aztec Soil Knowledge: Classes, Management, and Ecology." In *Footprints in the Soil: People and Ideas in Soil History*, edited by Benno P. Warkentin. Elsevier, 2006.

Withgott, Jay. "Is it 'So Long, Linnaeus'? In High-Stakes Name Game, Phylogenetic Nomenclature Puts Its Chips on the Table." *BioScience* 50, no. 8 (2000): 646–51.

Wolfe, Mikael. *Watering the Revolution: An Environmental and Technological History of Agrarian Reform in Mexico.* Duke University Press, 2017.

Wright, Angus. "Downslope and North: How Soil Degradation and Synthetic Pesticides Drove the Trajectory of Mexican Agriculture Through the Twentieth Century." In *A Land Between Waters: Environmental Histories of Modern Mexico*, edited by Christopher R. Boyer. University of Arizona Press, 2012.

Wulf, Andrea. *The Invention of Nature: Alexander von Humboldt's New World.* Alfred A. Knopf, 2015.

Zamudio Varela, Graciela. "El peregrinar de las colecciones de la expedición botánica de Sessé y Mociño." Transcript of presentation at the XV Congreso Mexicano de Botánico, October 15, 2001.

Zamudio Varela, Graciela. "Los artistas y las ilustraciones de la expedición científica a Nueva España." In *La Real Expedición Botánica a Nueva España*, coordinated by Jaime Labastida. UNAM, 2010.

Zárate, Julio. "El Estado de Veracruz en 1897." *Boletín de la Sociedad Mexicana de Geografía y Estadística* 4a época, vol. 4 (1897): 31–34.

INDEX

Page numbers in *italics* refer to figures.

ABOUT THE AUTHOR

Rick A. López is Anson D. Morse 1871 Professor of Latin American History and Environmental Studies at Amherst College. He is the author of *Crafting Mexico: Intellectuals, Artisans, and the State After the Revolution* and has published articles and essays on the history of nation formation, race, aesthetics, and the environment in Mexico and on the history the Latinx population in the United States.